Ecology, the Ascendent Perspective

Complexity in Ecological Systems Series

Complexity in Ecological Systems Series

T. F. H. Allen and David W. Roberts, Editors

Robert V. O'Neill, Adviser

Robert Rosen

Life Itself: A Comprehensive Inquiry Into the Nature,

Origin, and Fabrication of Life

Timothy F. H. Allen and Thomas W. Hoekstra

Toward a Unified Ecology

Ecology, the Ascendent Perspective

Robert E. Ulanowicz

Columbia University Press
New York

Columbia University Press gratefully acknowledges permission from the following publishers to reproduce or adapt the indicated figures:

Blackwell Publishers, Cambridge, MA	Figure 7.2	Page 126
Cambridge University Press, New York, NY	Figure 4.10	Page 89
Ecological Society of America, Washington, DC	Figure 4.2	Page 61
Elsevier Science, Inc., New York, NY	Figure 7.1	Page 122
Istituto Veneto di Scienze, Lettere e d'Arti, Venice	Figure 6.1	Page 116
Oikos, Lund	Figure 3.6	Page 51
Springer-Verlag New York, Inc., New York, NY	Figure 4.6	Page 74
	Figure 7.3	Page 128
	Figure 7.4	Page 129
	Figure 7.5	Page 130
	Figure 7.6	Page 131
	Figure 7.7	Page 134

Columbia University Press
Publishers Since 1893
New York   Chichester, West Sussex
Copyright © 1997 Columbia University Press
All rights reserved
Library of Congress Cataloging-in-Publication Data
Ulanowicz, Robert E.
     Ecology, the ascendent perspective / Robert E. Ulanowicz.
        p. cm. — (Complexity in ecological systems series)
     Includes bibliographical references and indexes.
     ISBN 978-0-231-10828-7 (cloth) — ISBN 978-0-231-10829-4 (pbk.)
     1. Ecology—Philosophy. 2. Biotic communities. 3. Causation.
     I. Title. II. Series.
     QH540.5.U42 1997
     577—DC21   96-37899
                                                              CIP

Casebound editions of Columbia University Press books are printed on permanent and durable acid-free paper.
Printed in the United States of America

To Edward and Mary . . .

for many years of love, sacrifice, and encouragement

CONTENTS

FOREWORD

The concept of ecosystem is now only half a century old, but it has changed its character a great deal during that time. By ecosystem we mean a view that explicitly includes the geophysical part of an ecological system inside the system. When Transeau (1926) worked on energy budgets using a cornfield and an orchard as his material systems of interest, he was implicitly employing a new conception of an ecological system. The emphasis on flux and process as opposed to place and organism was a departure from previous approaches. Without particularly emphasizing material flux, Tansley (1935) coined the term *ecosystem* as a modest solution to the terminological rats' nest of community ecology in his day. On the face of it, all Tansley said was that a reciprocity of influence exists between vegetation and physical environment and that the system should be structurally defined to include soil and air, as well as biota. If Transeau was explicit about a process-oriented view, Tansley cast the ecosystem as a structural entity, a set of biological and physical components. Tansley appears not to have anticipated where his term and concept was leading, but once he had juxtaposed the parts, the approach pioneered by Transeau followed naturally. With local environs included in the system, the devices of choice become mass balance, fluxes and budgets for minerals and energy.

Less than a decade later Lindeman's (1942) paper indicated the distinctive path the ecosystem concept was to follow. He still focused on species as his compartments, but the whole system was characterized as a system with energy flowing through it. The concept of ecosystem matured in the following decades, particularly as digital computers allowed the exploration of assumptions about energy and mineral fluxes and their rates. Odum's (1969) paper in *Science* suggested a set of principles for the ecosystem that are still being explored. Not only did energy and mineral flux serve as a descriptive device, but a new set of properties pertaining to overall system organization emerged in Odum's synthesis. Synthetic as Odum's ideas were, the principles still involved various types of connection and a wide variety of parts that might each require its own currency. Odum's principles might be characterized as a collection. The search was on for a new set of principles that would need to be connected out of necessity. Further developments in computational power allowed exploration of Odum's questions, as particular ecosystems were quantified and calibrated. The original conceptions of the ecosystem turned toward flux, and after half a century we are finally approaching an account of the general properties of ecosystems that uses the common currency of thermodynamics. A champion in this last struggle has been Bob Ulanowicz, and this book is his most complete and accessible statement on the subject.

The Complexity in Ecological Systems series is a natural place for Ulanowicz's seminal work. The present volume is that clear expression of ecosystem thermodynamics for which we have all been waiting. Robert Rosen, the author of *Life Itself*, was the first book in the series on complexity. Although Bob Rosen does not focus on ecosystems per se, many of us who were ecosystem ecologists cited him regularly. Ecosystem thermodynamics as a concept uses hierarchical thinking, and one of the hallmarks of complexity is hierarchical structure. The second book in the series, *Toward a Unified Ecology*, by Tim Allen and Tom Hoekstra, takes an explicitly hierarchical position. It examines half a dozen levels of organization, naming each one an ecological criterion. The ecosystem is one of those criteria. Allen and Hoekstra invite subsequent volumes in the series to amplify any one of those criteria or any combination of them. Ulanowicz's book takes the reader into an elaboration of the ecosystem criterion far beyond anything in the series' second volume.

The emerging science of complexity appears to include interwoven strands, and the notion of emergence and energetics is one of them. Ulanowicz gives us an original, powerful way to look at ecosystems. He also opens up a new set of principles whereby we can deal with complex systems through a small set of signatures. It is reassuring that Bob Ulanowicz's ideas are clearly related to extant notions of complexity and indeed may be an expression of an isomorphy. But of that time will tell. With the publication of this volume, ecosystem science secures new ground, and it takes nothing from Ulanowicz's achievement that it relates so well to previous work in ecosystem analysis done by himself and others. The school of ecosystem thermodynamicists was viewed askance by many of us; we asked ourselves what they were talking about. The significance of this approach is finally coming together in intuitive terms while retaining a solid technical grounding. In this volume, Bob Ulanowicz gives us a fresh and exciting exploration of ecological complexity.

T. F. H. Allen and David W. Roberts

PREFACE

Throughout the history of ecology, three metaphors have dominated dis-
cussions about community behavior: the ecosystem as (1) machine, (2)
organism, or (3) stochastic process. The keyword here is *or*, for it is
commonly accepted that the three models are separate and incommen-
surate. It should be easy, then, to imagine my excitement when, almost
twenty years ago, I chanced upon a single mathematical statement that
appeared to pertain to all three images.

The flush of excitement soon gave way to caution, however, because
current wisdom holds that only through the mechanical analogy may one
address cause and effect in a scientifically credible way. (True enough,
in the last century and a half there have been numerous attempts to integrate chance into mechanics, but I shall argue in this book that these
efforts have been either inappropriate or too narrow to adequately illu-
minate the dynamics of living systems.) As regards addressing causality
within the context of the organism, it seemed as though the obvious
directionality and apparent goal-seeking behavior of organisms led the
discussion inevitably to teleology, which remains anathema to most
biologists. Prudence seemed to dictate that I retreat into the unassailable
refuge of phenomenology, and so I chose *Ecosystems Phenomenology* as
the subtitle for my first book, *Growth and Development*, on the subject
of succession in ecosystems. It was a quantitative exposition of fundamental systems behavior with little reference to eliciting causes.

Over the subsequent decade it has become obvious, however, that
Growth and Development has had minimal impact on the dialogue
among ecologists over the nature of ecosystems. The relative lack of

response to the indices I defined and elaborated in my first book owes much to the mathematical way in which they were presented. I had been trained as an engineer to regard equations as fundamental and prose as subsidiary. It was an attitude that badly served my efforts to communicate to fellow ecologists. Colleagues, such as Robert O'Neill, and friends, such as Thomas Nadeau and my own wife, Marijka, encouraged me to rework my presentation to make it reasonably free of explicit mathematics.

In addition, I discovered to my chagrin that phenomenology is held in distinctly low regard by biologists—unlike its status in the physical sciences, where thermodynamics reigns supreme. Despite recent efforts by a few to rehabilitate empiricism in ecology, most ecologists are satisfied with nothing less than a full explanation of events. It gradually became obvious that I would have to face the issue of causality head-on and somehow summon the fortitude to thread my way "deftly through a conceptual mine field planted with vitalism and teleology," as I described the prospect in *Growth and Development* (Ulanowicz 1986a:4–5).

Fortunately, one gets by "with a little help from one's friends," as the saying goes. For example, organicism cannot be considered without somehow making reference to Aristotle. I was lucky while still in high school to have been nudged toward Aristotle by one of my mentors, Francis Stafford. More recently, Robert Rosen (author of the initial volume of this series) graciously accepted my invitation to participate in a conference I was organizing on the subject of theoretical ecology in biological oceanography. His cogent arguments as to why science should rehabilitate the notions of formal and final causalities marked the highlight of that conference and a turning point in my own thinking.

I soon began to write about the development of ecosystems in Aristotelian terms. A few years later I received from Copenhagen a small gift from Arne Feimuth Petersen, a gentleman I have never met. It was a brief monograph entitled *A World of Propensities*, written by Karl Popper. Arne complimented me on my Aristotelian narratives but let me know that he thought Popper said it all better. I wrote and thanked Petersen for his gift and promptly shelved it in my bookcase, where it remained untouched for the next six months. I ignored the book because I mistakenly associated Popper solely with logical positivism, which is specifically tailored for the reductionistic approach to science. It was with astonishment, therefore, when I finally read the essays, that I immediately realized how right Petersen had been. Popper does indeed provide the missing link capable of bringing the mechanical, the organismal, and the stochastic into a unified vision of natural phenomena.

Soon after I read Popper, it was most fortuitous that Henry Rosemont and Kenneth Tenore began organizing a graduate course on the philosophy and ethics of science to be given at our remote marine station. These exciting classes provided me with a welcome respite from the grind of "contract science" and provided much of what had been missing from my meager background in the history and philosophy of science.

I no longer had an excuse for delaying the rewriting of my description of ecosystem behavior that I had been encouraged to undertake. Which is not to say it was an easy task—writing rarely comes easily to one trained as an engineer. Once again, I was most fortunate in the outpouring of help I received from friends, colleagues, and even, in some cases, strangers (whom I now count among my friends). I am especially grateful to David Depew, who provided me with a line-by-line critique of my first draft, and offered me a multitude of new insights and references. Several others had the patience to read the entire manuscript and send their detailed comments; they include (in alphabetical order) Sabine Brauckmann, Robert Christian, Donald DeAngelis, James Kay, Stuart Pimm, Eric Schneider, and Rod Swenson. Still others were kind enough to read all or parts of the manuscript and offer me some suggestions for improvement: my thanks to Luis Abarca, Eduardo Cesarman, Jane Comiskey, Kevin DeLaplante, Martin Fleming, William Loftus, Koichiro Matsuno, Robert O'Neill, Claudia Pahl-Wostl, Stanley Salthe, and Laura Westra. I confess that I did not always heed the advice offered: several friends actively disagreed with some positions that appear in this volume.

My secretary, Jeri Pharis, showed enormous patience while retyping the manuscript many times. She also brought to the job her considerable experience as a librarian, and was most helpful in putting my bibliographic materials in good order. Frances Younger offered numerous suggestions for changes as she executed the illustrations. I also want to thank Columbia editors Timothy Allen and Edward Lugenbeel: Tim made several major suggestions for additions that appear in the final version, and Ed went out of his way to let me know how important the manuscript was to his press. As I had been unaccustomed to such encouragement, it became a genuine pleasure to do business with these gentlemen. Julia McVaugh was most considerate in her thorough and helpful editing of the initial copy.

In the end one looks to one's family for the closest support. My sister, Nancy, and her husband, Robert Konkol, kindly put their beach home at my disposal while I was writing; it provided me with a refuge from the office telephone and other constant interruptions. My parents, Edward and Mary, continued to offer me their encouragement, and my

Ecology, the Ascendent Perspective

1

ECOLOGY, THE EXCEPTIONAL SCIENCE

1.1   Malaise?

"Ecology is a sick science!" opined the renowned ecologist Daniel Simberloff (pers. comm.).[1] Written more than two decades ago, Simberloff's shocking exclamation continues to reverberate in the minds of many ecologists today. It is apparent to most who make their living through ecology that the direction of their discipline has not run parallel to that of other sciences.

But "sick"? Simberloff's assertion must seem doubly confusing to the avocational ecologist, who is inclined to identify "ecology" as the watchword of the environmental movement—which today is anything but sick. Never has consciousness of the environment been higher among the peoples of the West, and it is growing steadily among those of the world at large. Virtually every new enterprise is closely scrutinized for its potential consequences on the natural world. With the appearance of the "Green" parties of Europe, concern over the environment now has a formal presence on the political scene. Albert Gore became candidate for Vice President of the United States largely due to his call for action on environmental problems (Gore 1992).

It is true that some concepts arising from ecology have achieved widespread currency. What contemporary elementary school student is

---

1. This statement by Simberloff appeared somewhere in the gray literature and was widely repeated along the grapevine. When I contacted him recently to obtain the source, he said he distinctly remembers writing it but cannot recall where.

not familiar with the notion of the food chain, or better still, the food-web? The growing awareness of the natural cycling of materials (and, to a lesser extent, of energy) has prompted widespread efforts to recycle the materials of daily living. The adage that in ecosystems "everything is connected to everything else" has resonated in New Age culture to spawn a number of latter-day Spinozas (e.g., Ferguson 1980; Swimme and Berry 1992). Ecology as advocacy is vibrant and alive as never before.

Whence, then, this feeling of unease or outright defensiveness on the part of professional ecologists? On a personal note, the crux of the matter was illustrated vividly to me by a chance encounter. I was speaking with a well-educated individual who had telephoned to solicit my opinion on a particular local environmental matter. About ten minutes into the conversation she inquired about my profession. When I said, "I am a theoretical ecologist," she immediately shot back, perplexed, "What's that?"

It would be easy to attribute this woman's surprise and puzzlement to lack of familiarity with the discipline. Unfortunately, however, much the same view is shared by any number of those who study ecology for a living. Most ecologists feel there is no core principle or dictum that defines the science as, say, Newton's laws of motion establish the field of mechanics, Maxwell's equations prescribe the behavior of electromagnetic fields, or Darwin's survival of the fittest animates the realm of evolutionary theory. Despite a number of interesting and widely popular observations on regularities in ecosystem behavior, the conventional wisdom remains that ecology lacks a central dogma. In the words of Gertrude Stein, when asked to comment on her home city of Oakland, "There's no there, there."

I wish to argue in the pages that follow that the notion that ecology lacks a coherent theoretical framework is quite simply mistaken. The corpus of a theoretical ecology has been evolving steadily over the past few decades. Little attention has been paid to these developments by those outside ecology, however, or even by most of those who set the trends for the frontline ecological journals, and the reason for this neglect is not obscure: *The emerging picture of ecosystem behavior does not resemble the worldview imparted by an extrapolation of conceptual trends established in other sciences.* To be sure, ecology will continue to be enriched by excellent contributions from evolutionary theory, ethology, mechanics, thermodynamics, and other scientific disciplines. It is apparent to many, however, that ecosystems behave in ways that are very different from the systems described by other sciences.

1.2 *Omne Machina Est?*

It is no understatement to say that the mechanical analog has a particularly widespread and firm hold over how we view and practice science. Describing phenomena through the metaphor of the machine is the accepted (and to most the only) way of doing business. Accordingly, the discovery of the "mechanism" that drives a particular macroscale phenomenon is the holy grail of most ecologists. In what I have described as "Oecologia ex machina," ecosystems scientists seek to portray an ecosystem as a grand machine whose working parts are its component populations and whose linkages are its trophic and physical transfers (Ulanowicz 1993). We employ differential equations or algorithmic statements to represent these interactions and then integrate these mathematical constructs to predict how the system as a whole will behave. That is, starting with the state of the system at time t_1, we assume that this formulation will determine the system state at time t_2.

Such "mechanical models" of single processes or isolated populations often capture the behaviors of ecosystem components with reasonable fidelity (Platt, Mann, and Ulanowicz 1981). A problem arises, however, when the component models are spliced together. The ability of these aggregated mechanical constructs, otherwise known as "ecosystems models," to predict the behavior of ensemble ecosystems is notoriously poor. At a meeting of the North American Chapter of the International Society for Ecosystem Modelling, the ecosystem modeler Charles Hall challenged an audience of some sixty to seventy colleagues to name a single ensemble model that had been successful in predicting whole system behavior. The responses were memorable only for being so few and so qualified. More recently, Peter Abrams (1996) has discussed how our attempts to model systems containing several self-adapting components (which ecosystems always include) invariably lack the ability to predict.

I do not intend to denigrate the contributions of ecological modelers toward understanding ecosystems (I myself am a modeler): quite often, insights into what seems counterintuitive behavior have emerged, and continue to emerge, from our efforts, and it is most illuminating and satisfying when they do. Nor is anyone foolish enough to deny that mechanisms can be found at all levels in ecology, or that the elucidation of such mechanisms probably will continue to occupy the attention of most ecologists in the foreseeable future. My aim here is to discourage the extrapolation from the particular successes we have achieved, by mechanistic or quasi-mechanistic approaches, to the assumption that all ecology eventually will yield a grand mechanical picture of the natural world. Such a hope, I believe, is fatuous and will not lead to an accurate portrayal of nature (any more than the contemporary world of quantum physics conforms to the classical mechanical portrait of dynamics).

1.3   Cracks in the Facade

Although physicists over time have come to appreciate the shortcomings of a strictly mechanical approach to natural phenomena, it seems that biologists are reluctant to accept this conclusion. In part their reluctance to loosen their grip on the mechanical worldview is due to their reverence for the originator of the idea of natural selection, Charles Darwin, who was strongly influenced by the Newtonian determinism that characterized thinking during his time. (So much was determinism part of Darwin's exposition that it took a long time for later Darwinists to acknowledge all the roles that chance plays in natural selection [Depew and Weber 1994].)

Even today we inherit a rather peculiar view about causality in nature. As biologists, we relegate the generation of cause to the netherworld of molecular phenomena; once at the scale of cells, organisms, and populations, however, we imagine ourselves again in Newton's realm of strict determinism. Hence, to follow the process of neo-Darwinian evolution we are continually forced to shift perspectives abruptly from the stochastic world of Ludwig Boltzmann, where new genetic combinations arise, to the deterministic arena of Isaac Newton, where only those organisms with the fittest genes can be counted on to survive. So accustomed are we to constantly changing perspective that we rarely pause to ask ourselves whether causality in nature is really that schizoid. Yet ecology often is called to task for not having assimilated more of this bipartite scenario.

One attempt to obviate this schizophrenia has been to push the machine analogy ever deeper into developmental biology. Thus some have begun to write about "molecular machines" (Schneider 1991a,b). Mechanism in ontogeny is best illustrated by the assumption that a strict mapping exists between the structure of the genome and its expressions in the resulting phenome—that is, certain genes and genetic combinations predictably give rise to particular traits in the developed organism. It was to test exactly this hypothesis that Sidney Brenner and colleagues undertook to map exhaustively the correspondences between genes and traits in a very elementary multicellular nematode. Over almost two decades, millions of dollars went to support the best minds and the most competent laboratory technicians as they catalogued the effects of pre-scribed genetic substitutions upon the adult nematode. Perhaps the most remarkable thing to emerge from this project was the courage of the pro-ject leader, who ultimately declared:

> An understanding of how the information encoded in the genes relates to the means by which cells assemble themselves into an organism . . . still remains elusive. . . . At the beginning it was said that the answer to the understanding of development was going to come from a knowledge of the molecular mechanisms of gene con-trol. . . . [But] the molecular mechanisms look boringly simple, and they do not tell us what we want to know. We have to try to discover *the principles of organization, how lots of things are put together in the same place.*
>
> (Lewin 1984:1327; italics mine)

One way of rephrasing Brenner's conclusion is that we need to go beyond strictly mechanical metaphors to discover how biological pattern emerges. As an ecologist, however, I would suggest that ontogeny does not provide the best arena in which to discover Brenner's "principles of organization." The usual fidelity inherent in the replication of clones would indicate that the role of any unspecified "organizing principle" would remain small relative to that of mechanistic performance. To see where the analogy to machines truly breaks down, it becomes necessary to change scale and consider systems in which mechanical controls are less dominant.

1.4   Ecology's Favored Position

Another noted developmental biologist, Gunther Stent, offered the following:

> Consider the establishment of ecological communities upon coloniza-tion of islands or the growth of secondary forests. Both of these exam-ples are regular phenomena in the sense that a more or less predictable ecological structure arises via a stereotypic pattern of intermediate steps, in which the relative abundances of various types of flora and fauna follow a well-defined sequence. The regularity of these phe-nomena is obviously not the consequence of an ecological program encoded in the genomes of the participating taxa.
>
> (Lewin 1984:1328)

Stent suggested ecology as the ideal domain in which to pursue the study of organizational principles. I could not agree more. In ontogeny organizational influences per se are overshadowed by the mechanisms of transcription from genome to phenome. At the other end of the spectrum, human sciences—such as economics, sociology, or anthropology—involve the explicit exercise of volition. It clouds the issue to search for the most rudimentary nonmechanical organizing agencies when such higher-level complications exist. Ecology occupies the propitious middle ground. Here, as Stent suggested, it should still be possible to observe ecological organization as it emerges in systems relatively unaffected (until recently) by human activity.

Indeed, ecology may well provide a *preferred* theater in which to search for principles that might offer very broad implications for science in general. If we loosen the grip of our prejudice in favor of mechanism as the great principle, we see in this thought the first inkling that ecology, the sick discipline, could in fact become the key to a radical leap in scientific thought. A new perspective on how things happen in the ecological world might conceivably break the conceptual logjams that currently hinder progress in understanding evolutionary phenomena, developmental biology, the rest of the life sciences, and, conceivably, even physics.

1.5   Retrogression or Progress?

By this time readers may begin to have misgivings about the direction this discourse is taking. Did not the Newtonian revolution occur specifically to banish the transcendental and the metaphysical from scientific narration? Are we suddenly to eschew attitudes that have fueled the enormous advances in human welfare over the past three centuries? Is the mysterious the only alternative to mechanistic principles? These are legitimate questions.

The anxious should remain assured, however, that this work will remain within the bounds of natural phenomena. Only briefly in the final chapter will I digress to pursue the implications of ecological thought for broader, nonscientific human pursuits. There I argue that the mechanistic worldview, when followed too strictly, leads only to hypotheses that are far more metaphysical and transcendental than the very notions that Newtonianism sought to banish: witness Francis Crick's opinion (1982) that DNA is far too marvelous to have earthly origins, and so must have been seeded by extraterrestrial beings, or Richard Dawkins's suggestion

(1976) that genes are endowed with selfishness. By contrast, the ostensible mysteries of ecology will reveal themselves as fully in tune with what is truly natural.

As to suspicions that supplanting Newtonian assumptions on natural causality will somehow jeopardize both the scientific project and human welfare, let them be put to rest here and now. Many people today are concerned that specialists, often with the best of intentions, attempt to benefit society by applying their techniques with regard for only the most immediate consequences of their actions. Thus mechanical solutions to problems are proposed without regard for their consequences in the broader context, which might be supplied by the ecological perspective. For example, most genetic research is focused narrowly upon the creation of novel organisms designed to serve specific ends. Rarely is equal effort made to elaborate the ecological context in which such creations must persist, or to define the impacts that the new organism and its environment will have on each other. Ultimately, these are the larger questions concerning welfare.

Finally, building an ecological perspective is not to be confused with deconstructing the scientific project. As Brenner and Stent suggest, it is likely that only by adopting a new framework will further progress in understanding the living world be forged.

1.6 Themes to Come

Throughout this essay I will be mainly concerned with how events happen—i.e., with the description of causality in nature. As every astute ecologist knows, however, context should be a paramount feature of any description, and therefore I cannot simply launch into a description of ecological causality without first providing a brief preamble on the origins of the contemporary scientific attitude toward causality in living systems.

In the next chapter, I will outline some pertinent antecedents and consequences of the Newtonian revolution. As most readers already know, ecology is hardly the first discipline to challenge the Newtonian paradigm. I will delve briefly, for example, into the obstacles that thermodynamics posed to the Newtonian juggernaut. Whereas Newtonian mechanics was devoted entirely to quantifying the orderly phenomena of nature, thermodynamics was the first effort in science to juxtapose that order with the disorder seen in the world around us. Many elements of the thermodynamic endeavor are precursors to the ecologic perspective, so it will prove useful to spend time examining some key concepts in thermodynamics. Other challenges to Newtonian causality, such as arise from the discoveries of quantum theory and genetics, will be mentioned in passing. Eventually, we will encounter the dominant worldview in contemporary biology, the neo-Darwinian paradigm.

During this review I will gradually deconstruct the Newtonian worldview, which will lead to an acknowledgment of the role of chance and indeterminacy in shaping the living world around us. Followed to its extreme, this direction could lead to nihilism, nominalism, and absurdism; however, this is an extreme that I will abjure, in favor of the phenomenological notion that the natural world is a compromise between opposing tendencies toward order and disorder.

One of the most eloquent proponents of revising our views on causality was the late Karl Popper (1990), who followed in the footsteps of Charles Sanders Peirce when he declared his belief that the universe is causally "open." That is, he maintained that the deterministic realm where forces and laws prevail is but a small, almost vanishing subset of all real phenomena. The latter do not transpire entirely by chance, but are suffused with indeterminacies that confound efforts at deterministic prediction. Popper's is not the schizoid world of strict forces and stochastic probabilities, but rather a more encompassing one of conditional probabilities, or deep-seated *propensities* that are always influenced by their context or environment.

If Popper was a master at portraying the tradeoff between order and randomness, he unfortunately did not address the origins of the propensity of living systems toward order, nor did he attempt to quantify it. To approach the first issue I devote chapter 4 to the phenomenon of indirect mutualism in causal networks. Indirect mutualism bears on autocatalysis in chemical systems. It possesses distinctly nonmechanistic attributes, and it strongly affects system structure and function. My intent will be to cast the nonmechanical aspects of indirect mutualism as *agencies* imparting order to developing systems.

No endeavor can be considered scientific unless it possesses solid quantitative foundations. My earlier monograph, *Growth and Development: Ecosystems Phenomenology* (1986a), was devoted almost entirely to the derivation of indices and formulas intended to reflect the course of development in ecosystems. Central to that collection of measures was the term *ascendency*, an index meant to quantify the developmental status of a living community: *Ascendency combines the "size" or magnitude of system activity with the degree of coherency and organization among its component processes.*[2] Absent from this text are any but the simplest of mathematical formulas, allowing us to concentrate instead on the concepts behind measures that track the attributes of a living system as it grows, develops, and senesces. Ascendency and associated indices piece together various elements of the developmental puzzle, such as Popper's propensities, the effects of indirect mutualism, and the hierarchical character of causality.

The original formulation of ascendency was strongly Heraclitean in nature—that is, it relied entirely upon measurements of *process* rates (function) to the virtual exclusion of the *contents* of each component (structure). Only recently have contents or stocks been folded into the ascendency calculus in a rigorous and coherent way. In chapter 5 I will explore the ways in which stocks enter the picture, and will consider as well how the ascendency measure can be applied to spatial structures and temporal coherencies.

I do not pretend that the ascendency description of living systems that follows is wholly original or unique: there are numerous investigators groping at different members of the same elephant and describing what they feel in different terms. My hope, however, is that the following narrative describes the whole beast in more consistent fashion than can be found elsewhere. Chapter 6 relates the ascendency description to other, sometimes parallel theories now abroad. These connections are drawn with two goals in mind: I wish to acknowledge some of the sources for my own formulation, and at the same time I want to alert my readers to alternative approaches that they might wish to pursue.

It may seem presumptuous to tout the possible practical applications and philosophical ramifications of a theory that has yet to receive much press, but it is necessary to speculate in these directions lest the reader come away thinking that theorizing on the nature of causality in living systems is simply talk without consequences. Accordingly, chapter 7 is a brief summary of some applications for ascendency theory that have been attempted. These not only include the quantification

2. For those who may be curious, I have chosen the alternate spelling with an "e" to distinguish this complex ecological term from the common use of "ascendancy" to mean simply "dominance." The term as I use it here has a double meaning. In the conventional sense, a system with a greater ascendency has the capacity to dominate another system with less of the attribute. The second meaning comes from the root word "ascend" and so suggests the image of order "rising out" of chaos. For further discussion of the concept, see section 4.6 below.

of ecosystem status (e.g., response to perturbations, assessment of eco-system "health" or "integrity," comparison of trophic status, etc.) but pertain as well to problems outside ecology, such as evaluating the per-formance of neural networks, economic communities, and systems for distributed computation.

Finally, in chapter 8 I suggest that our perception of causes colors our approach to practical and philosophical issues. A holist, for example, would conduct research into finding a cure for cancer or AIDS in a very different way than would a reductionist. One's attitude toward the prac-tice of law and the administration of justice will likewise be affected by one's beliefs on how events originate. Opinions regarding individual rights vs. community responsibility, science vs. the arts and humanities, and, ultimately, science vs. religion, all turn on views of causality. It may be no exaggeration to claim that if society were to adopt the ecological perspective advocated here, it could begin to restructure itself in funda-mental ways. It would be a fitting way to enter the twenty-first century and a whole new millennium.

According to a Chinese proverb, "A journey of a thousand miles begins with a single step." Our first steps in reconsidering natural causality will be backward—back in time, that is, and away from ecology, to consider ideas and personalities that antedate the Newtonian revolution.

2

CAUSALITY IN THE AGE OF SCIENCE

There are several possibilities for dating exactly when the modern era, in which something called "Science" appears, actually began. Some point to the start of the seventeenth century, when Francis Bacon, Thomas Hobbes, and René Descartes urged emphasizing mechanical rather than teleological, or end-directed, causes. And certainly, the trial of Galileo was a landmark historical event. Without doubt, these influences prepared the way for a revolution. By common consent, however, the definitive moment came later in that century, *after* Isaac Newton had published his *Principia*.

To understand why the modern era turns upon Newton's work, it is necessary first to ask exactly what was overthrown by this revolution. Ultimately, it was the hegemony of the Church in matters secular as well as sacred that was being rejected. Galileo's confrontation with the Sacred Congregation had brought into question the role of the Church in interpreting secular phenomena. But it would have been entirely possible for the Church to have abandoned Ptolemy, embraced Kopernik, and still have conducted business much as before the confrontation—because Galileo and Robert Bellarmine (Galileo's inquisitor) were engaged in a dialogue mostly over *how*, rather than *why*, things happen.

2.1   Out of Antiquity

Quite inadvertently, Newton facilitated challenges to the prevailing belief about what causes things to happen that had reigned unquestioned since before the advent of Christianity. The Church's position on causality in nature was formulated largely by Thomas Aquinas, who in his turn had rewritten the philosophy of Aristotle to serve both secular and religious ends. Thomism, with its emphasis on God as final cause, was little more than Aristotle with a Christian slant, so that the Enlightenment crusade became directed in large measure against Thomism.

Aristotle's image of causality was more complicated than the one subsequently promulgated by the founders of the Enlightenment (Rosen 1985). He taught that a cause could take any of four essential forms: (1) material; (2) efficient, or mechanical; (3) formal; and (4) final. Any event in nature could have as its causes one or more of the four types. The textbook example for parsing causality into the four categories concerns the building of a house (table 2.1). Behind this process, the *material* causes are obviously the stone, mortar, wood, etc., that are incorporated into the structure, as well as the tools used to put these elements together. The workers whose labor brings the material elements together comprise the *efficient* cause. The *formal* cause behind the construction of a house is not as clear-cut as the first two. Aristotle posited abstract forms toward which developing entities naturally progressed. Thus he thought the form of the adult chicken was immanent in the fertilized egg; it was this endpoint that attracted all earlier forms of the growing chicken unto itself. This notion does not translate well outside the realm of ontogeny. The closest one can come to the formal cause for building a house, for example, is the image of the completed house in the mind of the architect. Usually, this image takes on material reality as a set of blueprints that orders the construction of the building.

Unfortunately, the image in the architect's mind is mostly determined by the purposes of those who are to live in the house. But these purposes

TABLE 2.1
Aristotelian Causal Typology as Applied to
Two Examples of Processes

Causal Type	Example 1: House Building	Example 2: Military Battle
Material	Stone, bricks, mortar, lumber, etc.	Guns, swords, tanks, ordnance, etc.
Efficient	Laborers	Soldiers
Formal	Blueprints	Juxtaposition of armies, lay of the land
Final	Need for housing	Social, political, and economic conflicts

are themselves what are identified as the *final* causes for building the structure. Hence, the house example too easily leads to the conflation of formal and final causes. I have suggested as an alternative example a military battle, which, despite its unsavory image, nonetheless provides a more appropriate and revealing example of formal cause (Ulanowicz 1995a). The material causes of a battle are the weapons and ordnance that individual soldiers use against their enemies. Those soldiers, in turn, are the efficient causes, as it is they who actually swing the sword, or pull the trigger, to inflict unspeakable harm upon each other. In the end, the armies were set against each other for reasons that were economic, social, and/or political in nature; together they provide the final cause or ultimate context in which the battle is waged. It is the officers directing the battle who concern themselves with the formal elements, such as the juxtaposition of their armies vis-à-vis the enemy in the context of the physical landscape. It is these latter forms that impart shape to the battle.

The example of a battle also serves to highlight the hierarchical nature of Aristotelian causality. All considerations of political or military rank aside, soldiers, officers, and heads of state all participate in the battle *at different scales*. It is the officer whose scale of involvement is most commensurate with that of the battle itself. In comparison, the individual soldier usually affects only a subfield of the overall action, whereas the head of state influences events that extend well beyond the time and place of battle. The intrinsic hierarchical nature of Aristotelian causalities was irreconcilable with the reductionistic and universal picture that early Newtonians sought to portray (David Depew, pers. comm.).

On this view, the very act of distinguishing four causes leads us to see that formal cause acts most frequently at the "focal" level of observation. Efficient causes tend to exert their influence over only a small subfield, although their effects can be propagated up the scale of action, while the entire scenario transpires under constraints set by the final agents. Thus, three contiguous levels of observation constitute the fundamental triad of causality, all three elements of which should be apparent to the observer of any physical event (Salthe 1993). It is normally (but not universally—e.g., Allen and Starr 1982) assumed that events at any hierarchical level are contingent upon (but not necessarily determined by) material elements at lower levels.

As Aquinas demonstrated, the purely secular views of Aristotle on causality can be adapted to numinous interpretation. It was probably this insistence, combined with the anticlerical temper of the seventeenth century, that led some philosophers of that time to play down all those elements in Aristotelian thought that most easily connect with the spiritual realm, and to concentrate instead on the more palpably physical aspects of reality. Thus it was that Thomas Hobbes, for one, arose to declare that *all* reality is in essence material. Hobbes even went so far as to suggest material interpretations of God and the human soul.

If Hobbes was the champion of material causality, it was Descartes who placed overwhelming emphasis upon the mechanical aspects of nature. At a time when town-hall towers and the drawing rooms of aristocrats were increasingly being filled with intriguing new machinery, it became inevitable that someone eventually should press to the limit the manifold analogies between natural events and fabricated processes.

By the latter part of the seventeenth century, intellectuals were growing increasingly predisposed to think mainly in terms of the material and the mechanical. Missing, however, was any forceful example of the dominance of these two causes in eliciting natural events. This missing ingredient was accidentally supplied to an awaiting audience by a very unlikely individual—Isaac Newton of Cambridge.[1]

2.2   The Newtonian Accident

By 1684 Isaac Newton's career was already going quite well at Cambridge, where his works on calculus and optics had won him appointment as Lucasian Professor (Westfall 1993). He had burst forth on the international scene in 1672 when his treatise on colors was published in *Philosophical Transactions*. The paper had attracted lavish praise from all over Europe—in fact from all quarters, save one. Robert Hooke considered himself the reigning authority on colors. He wrote Newton a lengthy critique written in a highly condescending tone. Newton waited four months to reply to Hooke, and then unleashed a reply that was "viciously insulting—a paper filled with hatred and rage" (Westfall 1993:93). This conflict prepared the stage for a monumental event that was to occur twelve years later.

In January 1684 Sir Christopher Wren (architect of St. Paul's Cathedral) and Edmund Halley (the astronomer who first located the comet bearing his name) met with Hooke in Oxford during a session of the Royal Society. Wren and Halley were both interested in establishing a rigorous connection between the inverse-square law of attraction and the elliptical shape of planetary orbits. When they inquired of Hooke whether such a connection was possible, Hooke told them he had already completed the

---

1. I am indebted to Robert Artigiani for pointing my attention to Newton's actual role in the revolution that bears his name.

demonstration but that he intended to keep the proof secret until others, by failing to solve the problem, had learned to value it (Westfall 1983).

Wren and Halley evidently were dissatisfied with Hooke's coyness, for in August of the same year, when Halley was in Cambridge, he sought out Newton in order to pose the problem to Hooke's rival. Newton told Halley that he too had already solved the problem but had mislaid the proof. These two cryptic claims to proof must indeed have perplexed Halley and Wren, for when Halley told Wren of Newton's reply, Wren decided to call the bluff of the two enemies by publicly offering an antique book worth forty shillings as a prize to the individual who could provide him a proof within two months.

Newton was distraught when Halley told him of Hooke's claim to a proof. Apparently, Newton did not lie when he claimed that he had already proved the connection, for a copy of a proof that antedates Halley's visit has indeed been found among Newton's papers. Being cautious in the company of someone who communicated with Hooke, Newton probably feigned having misplaced the proof. We can only imagine his horror, then, when he looked up his demonstration only to discover that it was deeply flawed. The thought of leaving the field open to Hooke drove Newton to near-panic. He abandoned all other ongoing projects to rush into seclusion and attempt a rigorous exposition. Once thus engaged, the positive rewards of the creative process seem to have drawn him ever deeper into the project. He virtually disappeared from society until the spring of 1686, at which time he emerged, on the brink of mental and physical exhaustion, with three completed volumes of the *Principia* in hand.

Most important to our theme on causality is exactly what Newton abandoned when he was seized by a fit of jealousy and anxiety over Hooke's claim. During the decade preceding Halley's visit, Newton had directed his energies primarily toward the study of theology and alchemy (Westfall 1993). He was a strongly religious man, albeit one who harbored secret, heterodox beliefs. His works just prior to the *Principia* were suffused with references to religious themes. Caught up, however, in the throes of creative splendor, and pressed to finish his ideas on dynamics before Hooke could preempt him, Newton had time to write only the barest account of the technical elements in his theory. There would be time enough later, or so he must have thought, to replenish his account with appropriate theological references. In the context of growing mechanistic materialism, however, this oversight was to have large consequences.

It is not difficult to imagine the delight of Edmund Halley, who by all accounts was probably a closet materialist, when he first leafed through the volumes of *Principia*. Here for the first time was a rigorous, minimalist description of the movements of the heavenly spheres, without any recourse whatsoever to supernatural agencies. Halley immediately set about to help Newton publish his grand work as it stood. The books immediately caught the attention of the awaiting followers of Hobbes and Descartes. Newton quickly became a legend in his own time, and he remains one today: the founding father of a mechanical philosophy of nature that he in fact did not hold!

Fame has its price, and Newton must have perceived that rewriting the *Principia* in his usual theological/alchemistic manner hardly would have enhanced his image in the eyes of his new supporters. Neither of the subsequent editions published in 1713 and 1726 was rewritten in his earlier elaborate style, although a General Scholium addressing numinous concerns was appended to the second edition. At some point Newton did begin his unabridged version, but he soon abandoned the effort. Pieces of the aborted revision lay undiscovered until early in this century.

In the wake of Newton's monumental achievement, the window on what constituted a legitimate scientific explanation quickly narrowed to admit only material and mechanical (efficient) causes. The project of elaborating the "universal clockwork" grew unimpeded throughout the eighteenth century. It is difficult to exaggerate the pressure that Enlightenment scientists placed upon each other to make all scientific descriptions conform to the Newtonian mold. Enthusiasm for Newtonian explanation reached its apogee early in the following century when Pierre Laplace (1814) exulted over the unlimited horizons of the mechanical worldview: any "demon" or angel, he exclaimed, that had a knowledge of the positions and momenta of all particles in the universe at a single instant, could invoke Newtonian dynamics to predict all future events and/or hindcast all of history.

My purpose in belaboring the events leading up to Newtonianism is to show how from the very beginning there were strong inconsistencies between how Newtonianism paints the whole world of action and the particular events that gave birth to the Newtonian worldview. Our Newtonian heritage inclines us to imagine the revolution in very material and mechanical terms. As material instruments there remain the original manuscripts and early copies of the *Principia*. Newton himself was the efficient agent who put together his works. The strict reductionists might even speculate about the neuronal connections in Newton's brain—what lay behind them, and what events they subsequently set into motion. We are wont to portray Newton as the first man of the modern era, an individual agent whose singular genius changed the course of history.

We like heroes, and Newton is unquestionably one. But Frank Edward Manuel (1968) suggests, and Richard Westfall's account supports his opinion, that Newton's own worldview made him a better candidate for the last medieval man than for the first modern man. Had his pursuit of science continued in the direction that Newton, the Arian and alchemist, intended, the Newtonian legacy would loom nowhere near as large as it does today.

It happened that the *Principia* told the right story at the right time— exactly the sort of story the materialists were poised to receive and loudly proclaim. The accidental form of Newton's account was as much responsible for the monumental surge of Enlightenment thought as was his undeniable genius in creating modern dynamics. The social milieu obviously had its reciprocal effects upon Newton—witness his obvious reluctance to make his intended addenda to the *Principia*. This is not to say that just any new breakthrough in dynamics would have served to precipitate the revolution: it is unlikely that a treatise by Hooke would have sufficed. But it is the larger social nexus that appears as the primary agency in fashioning the subsequent fame and meaning of the *Principia*.

2.3   From Newton to Darwin

Like wildfire, the Newtonian way of regarding natural phenomena invaded and transformed all the physical sciences (Depew and Weber 1994). Antoine-Laurent Lavoisier and John Dalton brought Newtonian thinking to bear on the problem of chemical affinities. Benjamin Franklin and James Clerk Maxwell applied it to electrical and magnetic phenomena. Charles Lyell applied it to geology. But for a while a firebreak appeared to halt the advance of Newtonianism into biology. It was Charles Darwin who was to carry the Newtonian approach across this final Rubicon.

Darwin is not usually thought of as a disciple of Newton. Darwin's own account credits Thomas Malthus and Adam Smith for his greatest inspiration—but, as David Depew and Bruce Weber (1994) argue, these thinkers were only a slight detour through which Newtonian thinking traveled to reach Darwin.

Malthus, the reader may recall, wrote that a population has an innate potential to increase at a geometric rate. The population's food supply, meanwhile, grows at best in arithmetic proportions, so catastrophe is sure to ensue. The Newtonian idea here is that the population follows the geometric pathway of its own "inertia" in the absence of any intervening "forces"—that is, a population will follow its inertial trajectory until food scarcity acts in the guise of a "force" to brake that increase. This is a scenario right out of Newton. It should be remembered that Newton's first law was quite revolutionary: it said that a body moving in a straight line would continue to do so until acted upon by external forces. Prior to Newton no special status had been accorded straight-line motion. In fact, Aristotle had argued that circular motion (e.g., the Ptolemaic planetary orbits) was the most natural—the circle being the most perfect geometric form. But Newton's radical idea was that planets moving in isolation would cut a straight path through space. The arcs we see them follow are the results of other forces (exerted by other heavenly bodies) acting at every instant to bend each trajectory away from a straight line.

From Adam Smith Darwin absorbed the notion of atomism. That is, an economic community is best considered as a collection of independent, self-interested agents, who, if they all were left free to follow their own pursuits, would arrive at a balanced and most propitious configuration. Thus, Darwin came to regard populations on the evolutionary stage as acting solely for their own behalf and *independently of surrounding populations*. As with a collection of Newtonian bodies, one may treat each component in isolation and describe the aggregate behavior via addition or subtraction.

It is important to note that Malthus and Smith transformed Darwin's thinking by removing most biological causality from the inner world of organismal development and transformation (thereby consciously distancing his ideas from those of Aristotle) to the external world of variational and selectional forces (Depew and Weber 1994). The switch was a crucial step in Darwin's quest to become the Newton of biology (Schweber 1979).

Virtually all historians agree that the eighteenth century saw the unimpeded spread of the Newtonian viewpoint into all branches of science (save biology). As we have seen with Adam Smith, the analysis reached even into the social and political spheres, where society came to be regarded as one grand machine. It was the highest accolade one could bestow during those heady years of the Enlightenment to call someone a mechanic. Garry Wills (1978) elaborates how the founding fathers of the American Revolution shared a common image of society and government as machinery, to the extent that one could say they "invented" America.

Just as Newton postulated that any force always elicits an equal and opposing force, the Romantic movement arose in the arts and literature to counter what its participants regarded as the overbearing determinism of the Newtonian outlook. But Romanticism made few significant inroads into the sciences. Not so gradually, a chasm was forming that separated the arts and literature from the sciences to an extent much greater than ever before.

2.4   Thermodynamics: The Challenge by Engineers

Most historians of science note that the nineteenth century was dominated as well by the inexorable march of Newtonianism. We have seen, for example, how Darwin carried its banner into the redoubts of biology. Indeed, by most accounts the Newtonian worldview remained intact until the turn of the twentieth century brought with it the advent of relativity theory and quantum physics. Some narrators make passing mention of the appearance of thermodynamics early in the 1800s and of how it elicited misgivings on the part of some concerning the universality of Newton's theories—but this challenge, if mentioned at all, is usually considered a mere hiccough in the proliferation of Newtonian thought. Besides, the accepted view is that thermodynamics eventually was reconciled with the mainstream of modern thought through the statistical mechanics of Josiah Willard Gibbs and Ludwig von Boltzmann. Yet apparently Boltzmann himself did not feel that way: by all accounts, he felt alienated from the prevailing attitudes of his peers and eventually committed suicide. It may therefore be informative to reconsider the relationship of thermodynamics to mainstream Newtonian thought. In addition, some of the concepts developed in thermodynamics will later prove crucial to piecing together an ecological perspective on the world.

Before delving into thermodynamics, it is useful to pause and summarize the foundations that underlay the Newtonian construct. Thus far I have emphasized only its identification of mechanical and material causes to the exclusion of all others. I also briefly mentioned the notion of decomposability. Depew and Weber (1994:92) cite four conceptual presuppositions that define the Newtonian approach:

1. Newtonian systems are *deterministic*. Given the initial position of any entity in the system, a set of forces operating on it, and stable closure conditions, every subsequent position of each particle or entity in the system is in principle specifiable and predictable. This is another way of saying that mechanical causes are everywhere ascendant.
2. Newtonian systems are *closed*. They admit of no outside influences other than those prescribed as forces by Newton's theory.
3. Newtonian systems are *reversible*. The laws specifying motion can be calculated in both temporal directions. There is no inherent arrow of time in a Newtonian system.

4. Newtonian systems are strongly decomposable, or *atomistic*. Reversibility presupposes that larger units must be regarded as decomposable aggregates of stable least units—that what can be built up can be taken apart again. Increments of the variables of the theory can be measured by addition and subtraction.

To this list we should add:

5. Newtonian laws are *universal*. They apply everywhere, at all times, and over all scales.

As noted explicitly in item 4, the five attributes are not entirely independent of each other (i.e., the Newtonian vision itself is not strongly decomposable).

Against this background let us now examine the observations of the French military engineer Sadi Carnot (1824). Carnot was interested in improving the design and operating characteristics of early steam engines, used then principally to pump water from mines. He set out to learn how much water could be pumped by machines that operated over different cycles of compression, heating, expansion, and cooling. After numerous preliminary trials, he came to the most interesting conclusion that "it is impossible to construct a device that does nothing except cool one body at a low temperature and heat another at a high temperature" (Tribus 1961:149). In subsequent trials he turned his attention to combinations of possible cycles, in search of the most efficient. By applying the principle he had just formulated, he was able to demonstrate that there is a particular cycle that is more efficient (in terms of converting thermal energy into work) than any other imaginable. To this day, the performance of working gas machines is still compared to the efficiency of this "Carnot cycle."

In the years that followed, Carnot's principle appeared in any number of equivalent forms. Rudolf Clausius, for example, restated Carnot's observation thus: "Heat cannot of itself flow from a colder body to a warmer one." Of course, it is easy to imagine heat flowing of itself from a warmer body to a colder body. This happens all the time; it is the natural way for heat to behave in the absence of external influences, to state it in terms that Newton might have used. But Clausius's prohibition leads to a conflict with the Newtonian requirement that all systems are reversible. If we see heat flow from a hotter to a colder body, why, if we wait long enough, do we not see it flow the other way? Put another way, suppose we agree with Laplace, who said that if we were able to know the positions and momenta of all the particles that comprise a system, we could predict all future states of that system. Combining the atomic hypothesis, formulated by chemists, with the decomposability of Newtonian descriptions, we should in principle be able to describe this system at the atomic level. At this scale we would see particles interacting with one another, and presumably we could describe them using Newtonian mechanics. If we reversed time for each individual interaction between particles, the reverse trajectories all would look wholly plausible. If we then combined all these plausible trajectories, the attribute of decomposability should yield for us a plausible prediction of system behavior. But that is not what would happen; instead, we would see heat flowing of its own volition from a colder to a hotter body, which, according to Clausius, *never* occurs. The inference seems to be that the Newtonian attribute of (1) predictability, (4) atomism, or (5) universality (or some combination of these three) conflicts with Carnot's phenomenological observations.

Actually, it is the property of (2) causal closure that is called most into question. But to see why, we must first consider how Newtonian dogma affected measurements of heat and work. Gottfried Wilhelm Leibniz's law of conservation of momentum proved to be a very powerful bookkeeping tool, and the equivalence of kinetic and potential energy followed by deduction. It was a magnificent tool, this conservation, and doubtless it strongly influenced the work of Joseph Black (1806) on calorimetry. Black studied the changes in temperature as heat was transferred from one substance to another. He concluded (without explanation) that various materials differ in their capacities for heat; heat can flow from one body to another, but its total amount *must remain constant* (Moore 1962). This idea of a conservative *caloric*, which Lavoisier listed in his table of chemical elements, soon was to be amended to account for latent heats of phase transitions and chemical reactions. But up to the *mid-nineteenth century* no serious attempt was made to fold the concept of caloric back into the engendering notion of mechanical work and energy. (Although Carnot must have realized that heat could be used to do work, that conversion did not enter his calculus. His focus was on calculating the mechanical energy of the gas at each phase in the cycle of machine operation.)

There were early indications that heat and mechanical energy might be interconvertible. For example, Benjamin Thompson suggested in 1798 that the heat arising during the boring of cannons was related to the mechanical work expended in the process. Furthermore, Dalton, in a corollary to his atomic hypothesis, provided an understanding of heat in terms of molecular motion. But it was the physician Julius Robert Mayer who, in 1842, while pondering the fate of food as the source of both heat in and work from the human body, finally provided the numerical equivalence between heat and work. This principle of interconversion was refined by James Joule, who in 1848 published data from very careful quantitative experiments on the interconversion of energy between its various forms. Hermann von Helmholz thereafter clearly and quantitatively declared the conservation of energy as a principle of universal validity—a fundamental law applicable to all natural phenomena (Moore 1962).

The equivalence of heat and work having been fully established, it then became possible to incorporate the conversion of heat and work into Carnot's dictum: In any real process it is impossible to convert a given amount of energy entirely into work. Some energy will always be left unavailable for work via any process. How much energy in a gas cannot be used for work was found to vary inversely with the absolute temperature of the gas—that is, a greater fraction of the energy in a hotter gas could be turned into work than it was possible to extract from a gas at a lower temperature. This relationship between available work and temperature gave rise (through an analytical process we will not discuss here) to the concept of the entropy of a substance. In the simplest terms possible, the entropy of a substance is the increment of energy that becomes *unavailable* to do work when the absolute temperature of the substance drops by one unit. The higher the entropy of a system with respect to its state at absolute zero (0° Kelvin), the less energy the system has available to do work. To quantify the amount of energy available to do work, the product of the absolute temperature times the entropy of the substance is subtracted from the total energy. The result is called either the Helmholz free energy, the Gibbs free energy, or the exergy (Evans 1969), depending upon how the total energy was reckoned.

Returning now to how Carnot's observation disrupts closure, it is evident that whenever we deal with real processes, something inevitably is lost. Although our bookkeeping on energy tells us that all the energy at the beginning of the process still exists when it is over, entropic considerations inform us that there truly is *less capacity for change* in the final condition than at the beginning. When trading in entropy and related variables, such as free energies, we are dealing in the realm of nonconserved quantities. Entropy, the Carnot maxim tells us, is always being created ex nihilo. Free energies, on the other hand, are continuously disappearing. If our focus is upon the true capacity for change embodied in a system, we must learn to speak in very nonconservative, and therefore non-Newtonian, terms.

From all this it follows that it is the free energy or the exergy, not the total energy, that is most informative in the examination of dissipative (that is, entropy-producing) systems. Ecosystems and organisms are systems of this sort. Tracking nonconserved quantities tells us most of what we want to know about how living systems are behaving. The fact that there is something more general called energy that is conserved, which is so important to physicists, fades into the background. Energy begins to seem less a concrete reality than an artifact of a constructivist bookkeeping system necessary for estimating quantities whose roles seem far more palpable and physically relevant (Reynolds and Perkins 1977).

It has always seemed to me passing strange that the law of conservation of energy should be the only law in all science to hold without exception, everywhere, all the time. I have often speculated that perhaps things were merely defined in exactly such a way that they would always balance. Science was on a roll, what with the discoveries of conservation of momentum and caloric. Newton had declared the world to be a closed system, so why not simply *define* energy to be conserved? Mayer and Joule conveniently came up with the numbers that would make it so. Physicists might argue that such musings border on the insane. The convenience and power that result from treating energy as a conserved quantity *in nearly isolated systems* render it almost impossible *not* to reify a conserved energy. I can respond only by saying that the *utility* of treating the capacity for change as a conserved quantity becomes increasingly less compelling the farther one is removed from isolated (and hence, rare) physical circumstances. Even in purely physical systems the case for total energy is not watertight—witness the considerable waste that power-generating technology incurred until quite recently, because power engineers were so set upon reckoning efficiencies according to absolute (conserved) energies instead of following the example of their colleague of yesteryear, Sadi Carnot, who addressed the (nonconservative) capacity for work (Ford, Rochlin, and Socolow 1975; Gaggiolo 1980; Hevert and Hevert 1980).

Most readers are probably familiar enough with thermodynamics to recognize that for the last few pages I have been discussing the first and second laws of thermodynamics. I have purposely refrained from calling them by name, in order to make a point. Carnot's discovery occurred almost thirty years *before* the conservation of work-energy was fully articulated. His principle allows us to define attributes fundamental to our understanding of dissipative processes, which happen to comprise most of the observable world. Nonetheless, it is made to sit in the back seat, so to speak: it is called the *second* law of thermodynamics. The conservation of energy, which was not expounded in its entirety until well *after* Carnot's observation, is nonetheless accorded the cardinal status of *first* law. I would suggest that this ranking owes to a tacit nineteenth-century consensus that the maintenance of a closed world was paramount to science. It is a construction—one that throws us off the right sense of proportion about nature.

No one is claiming that the first law of thermodynamics is in error. It is only that its *utility* is nowhere as universally great as most think it to be. In ecology, as in all other disciplines that treat dissipative systems, the first law is not violated, but it simply does not tell us very much that is interesting about how a system is behaving.

There is, indeed, a school that holds that biological phenomena in general are "autonomous" of all physical constraints such as those of thermodynamics (Varela 1979). Ecologists usually do not go that far, because the very definition of the ecosystems they are studying includes physical elements (Tansley 1935). As regards thermodynamics in ecology, there is in fact a growing recognition of the central role played by the second law and dissipative behavior. Even thermodynamicists themselves have begun to question the independence of the first law and are attempting instead to forge a unitary statement that incorporates both laws (Hatsopoulos and Keenan 1965). Some ecologists are keen on the potential that such a unified statement has for improving our description of ecosystem behavior (Schneider, E. D. and Kay 1994a,b). Others (Rod Swenson, pers. comm.) maintain that the two laws must remain separate: they argue that one cannot conceive of dissipation in the absence of conservation. Change and stasis are necessarily complementary and can no more be joined than can the dialectic between Parmenides and Heraclitus (stasis vs. change) be resolved. I shall return to this theme in chapter 6.

2.5   Statistical Mechanics: A Reconciliation?

If Carnot's discovery poses such a challenge to the Newtonian clockwork, why then do most writers ignore it and reckon the first cracks in the Newtonian facade from the advent of quantum physics? The answer is that it is commonly believed that the difficulties initially posed by thermodynamics were all adequately resolved by the appearance of statistical mechanics late in the nineteenth century.

In statistical mechanics one begins by assuming that gases are comprised of atomistic particles that move about according to the laws of classical mechanics. Since there is no way to treat analytically the immense numbers of particles in any visible volume of gas (treating three particles at one time is difficult enough!), *ignorance* about the details is side-stepped by assuming that the positions and momenta of the pointlike particles follow some statistical distributions. Through the application of the methods of statistics it is possible to show how the perfect gas laws arise. It is even possible to treat gases whose atoms have sizes and shapes of their own and interact weakly with nearby particles according to some simple force law. In this way one is able to describe, and to a degree predict, the deviations that many real gases exhibit from the perfect gas law (Chapman and Cowling 1961).

Of most interest, Boltzmann asked how any initial statistical distribution of particles would most likely evolve if left free of outside influences. He discovered a particular function of the particle probability distribution that always increased monotonically in time, and he identified this function with the entropy of the gas. It thus became conceivable that a collection of particles could act in Newtonian fashion at the microscopic scale but nevertheless exhibit irreversibility in the large. Carnot and Clausius, it seemed, had been brought within the orbit of Newton, and all was well again with the scientific world.

In almost any reconciliation, some questions are left unanswered and then ignored—and so it was with statistical mechanics. Probability theory and statistics, for example, had been created expressly to circumvent an observer's *ignorance* about detailed events that everyone assumed were amenable to classical deterministic mechanics, e.g., the throw of a die or the spinning of a roulette wheel. In the course of time, however, it became obvious that the same mathematics could be applied to events that were inherently stochastic (provided, of course, that one could conceive of such indeterminacy in a world of Newton's laws). It appeared that Boltzmann could conceive all too clearly of a universe that was stochastic and random at its core—and that possibility seems to have contributed to his despair. Since the advent of quantum physics, growing credibility has been accorded to indeterminacy over ignorance as the proper object of statistical considerations, although the controversy remains very much alive to this day (Prigogine and Stengers 1984).

On a personal note, I find this rush of consensus to consider the reconciliation complete a bit odd and premature. Odd, because the majority of those who are quick to drop the matter belong to the positivist school of doing science. Positivist doctrine bids us consider only hypotheses that are capable of being proved wrong—i.e., falsified. A hypothesis, once abroad, should be subjected to repeated attempts to falsify it under a myriad of different conditions. A failure of the hypothesis anywhere is grounds at the very least for amending it, and possibly for rejecting it entirely. As a community, however, we have acted in quite the reverse way with regard to the reconciliation provided by statistical mechanics. We were eager to hold on to the Newtonian perspective in spite of the thermodynamic challenge. We found only a very narrow set of conditions (rare gases, close to thermodynamic equilibrium) where the two were possibly in accord, yet the case was immediately closed. Newtonian mechanics and thermodynamics are now implicitly considered compatible over all conceivable circumstances. Such is the power and attraction of the Newtonian vision over our collective psyche, even when we think we have grown beyond it!

This last criticism is directed mainly at contemporary attitudes toward the compatibility of classical mechanics and thermodynamics. In fairness it should be added that during and long after Boltzmann's time the *only* systems to which thermodynamics could be applied were those in equilibrium. In thermodynamics "equilibrium" has a far narrower meaning than that of an unchanging balance. It may happen, for example, that several dissipative processes, such as diffusion and chemical reaction, occur at rates that just balance each other, so that an observer sees no change for quite a while; such a system is said to be at steady-state. But it is definitely not in thermodynamic equilibrium: only if no dissipation is occurring within the system—that is, if no entropy is being created—can it be regarded as at equilibrium. A simple thought experiment can be invoked to decide whether a system is only at steady-state or is at full thermodynamic equilibrium: One imagines first that the system in question is suddenly isolated; that is, all its exchanges with the outside universe are severed. If, after isolation, the system remains completely unchanged, then it was originally at thermodynamic equilibrium. If, on the contrary, *any* changes ensue, then it was at steady-state, but not at thermodynamic equilibrium (Ulanowicz 1986a).

It should be obvious from this test that very few of the systems we encounter are at thermodynamic equilibrium, and most certainly no living systems are among these. For the most part, equilibrium systems are merely imaginary devices used in thermodynamics to represent and demonstrate a particular point. Another common device is the reversible pathway. One imagines a system going from one (equilibrium) state to another via a succession of infinitely closely spaced intermediate equilibrium states—a succession that would take infinitely long to transpire. Obviously, equilibrium pathways are not real and can only be approximated, given enough time and patience. The key advantage in the concept of a reversible pathway derives from its name: it is *reversible*. It is accompanied by no dissipation. In fact, any equilibrium system may be divided in any fashion into component subsystems (it is atomistic). No macroscopic exchanges of material or energy are allowed, for they would engender dissipation. The system is effectively closed. Finally, reversible pathways can be described by what is called the "equation of state"—i.e., they are predictable. In short, systems at thermodynamic equilibrium satisfy all the Newtonian prerequisites. It is little wonder, then, that a perfect gas near equilibrium can be described by Newtonian microscopic variables.

The more important question, if we hope to apply thermodynamics to living systems, is how to describe systems that are *not* at equilibrium. Strictly speaking, thermodynamic state variables (such as temperature, pressure, or entropy) can be defined only at equilibrium. It may surprise many readers to learn that serious efforts were not mounted to extend thermodynamics beyond equilibrium descriptions until about a hundred years after Carnot published his findings. Unfortunately, the results of such efforts leave much to be desired.

2.6   Irreversible Thermodynamics: A Dubious Step in the Right Direction

Contemporary irreversible thermodynamics begins with the assumption that in a system near enough to equilibrium it is possible to define thermodynamic variables as scalar fields. That is, the spatial extent of the system is divided into an imaginary gridwork of sufficiently small cells. The cells are large enough to contain a macroscopic amount of material, but small enough so that the differences between a cell and its neighbors are very small. State variables are approximated by a spatial succession of cells, each at thermodynamic equilibrium. Temporal variations within any segment must remain very slow. Gradients in such state variables and fields are accompanied by measurable flows. For example, a spatial gradient in temperature is accompanied by a concomitant heat flux (thermal diffusion). A similar gradient in chemical free energy is linked to a diffusive material flux. The gradient need not always be situated in physical space: it may exist along some abstract dimension, such as the extent of reaction (i.e., how far a chemical reaction is from equilibrium).

Long ago, many of these physical flows were related to their accompanying gradients in phenomenological linear terms patterned after Newton's second law. It was Joseph Fourier who described the conduction of heat as varying in proportion to the negative gradient in temperature. Adolf Eugen Fick suggested that mass diffusion was proportional to the negative of the gradient in concentration of the substances in question. Thus it was that heat and mass were said to "flow downhill" along their gradients in temperature and concentration, respectively. The important point to notice is that such diffusive flows are spontaneous—that is, they will occur unless some opposing constraint keeps them from happening. When they do occur, moreover, they generate entropy, as do all spontaneous processes. The pioneers of "irreversible thermodynamics" astutely identified the magnitude of entropy generation as the important measure of activity. Unfortunately, they followed tradition in describing these irreversible events in the language of reversible mechanics. In mechanics, if something flows downhill, it is because of an eliciting force (gravity); hence, behind every observed spontaneous flow was postulated a conjugate thermodynamic "force." Thermodynamic forces were defined with dimensions such that the product of a force times its respective flow yields the entropy generation per unit time by that process. To calculate the entropy generation rate for an ensemble of processes, one multiplies each flow by its conjugate force and sums over all such products (Onsager 1931).

The linear form of Newtonian mechanics was copied in this analysis, so that near thermodynamic equilibrium a flow could be said to be proportional to the magnitude of its eliciting forces; the factor of proportionality was called a "phenomenological coefficient." The reader will notice the plural "forces" in the preceding statement: this was because forces were allowed to interfere with each other, giving rise to secondary effects called "coupling." For example, imposing a thermal gradient upon a homogeneous fluid or solid mixture results not only in a flow of heat through the medium, but in the differential migration of one or more of the species in the mixture. This coupling of mass diffusion with heat flux is called the Soret effect, named after its discoverer, Charles Soret. Conversely, the thermal conduction induced by a mass flux is called the Dufour effect, named for its discoverer, Louis Dufour (see Ulanowicz 1986a:23, example 2.3).

Lars Onsager (1931) was able to demonstrate that near equilibrium (that qualifying phrase again), the phenomenological coefficients that characterize coupling phenomena are symmetrical. That is, the coupling coefficient that relates the thermal gradient to the mass flow (the Soret effect) is equal in magnitude to that which links the gradient in chemical potential to the heat flux (the Dufour effect). One finds in Onsager's law of symmetry intimations of the Le Châtelier–Braun principle, which says that any perturbation to a factor contributing to equilibrium induces a compensating change in an opposing factor. Thus, a disturbance in a temperature distribution causes not only heat to flow, but mass as well. The symmetry in coupling coefficients ensures that the gradient in chemical potential that appears will be opposite to the perturbing temperature gradient.

Ilya Prigogine recognized the Le Châtelier–Braun principle in Onsager's law of symmetry and formulated a unifying and simplifying description of how such ensembles behave near equilibrium. He demonstrated that an arbitrary collection of processes sufficiently close to equilibrium "always operate in such a way that their effect is to lower the value of the [specific] entropy production per unit time" (Prigogine 1967:83). Like Hamilton's principle in classical mechanics, Prigogine's statement is variational: it identifies an objective function (the community entropy production) that tends toward an extreme value (a minimum in this case). Note, for later reference, that Prigogine's principle, whatever its limitations may be, pertains to the whole ensemble of processes. The various components of the whole system, the forces and fluxes, all *coevolve* so as to minimize the entropy production. In other words, the elements of nonliving, nonthinking processes behave like a *protocommunity*. This observation should give pause to those who are rigid in eschewing a legitimate place for telos or final cause in biology.

For a while, the formulations of Onsager and Prigogine aroused much optimism that thermodynamics soon would be successfully extended to encompass the larger world of nonequilibrium phenomena. In the end, however, things did not develop this way, although some useful concepts did arise from the endeavor. It is informative, accordingly, to discuss briefly some of the reasons why the effort lost momentum.

The problems with the early formulation of irreversible thermodynamics were both practical and conceptual. On the practical side, fundamental to any scientific endeavor is the necessity to identify and measure the principal elements in any narrative. In particular, we ask whether the force conjugate to each observed flow is always obvious. Can it be generalized in the sense required by Onsager's analysis? Can it be measured? The answer to all these questions is usually positive, so long as our observations are confined to simple substances near equilibrium. But matters quickly become equivocal as soon as the system in question does not satisfy these assumptions. As an example, consider foxes feeding on field mice. In a legitimate sense this transaction can be regarded as a flow of material and energy from the population of mice to that of the foxes. What, however, is the force behind the exchange? One's first inclination is to point to the energy gradient between the mice and the foxes. In the sense of the first law, there most likely will be more total energy in the mouse population than in the collection of foxes—so should we adopt the perspective of H. T. Odum (1960) and try to think of the mice as "forcing themselves" into the stomachs of the foxes? Perhaps instead we should recall how the concept of total energy does not necessarily capture the "capacity for change" in a system. Precisely to accommodate the second law, the thermodynamic force associated with mass diffusion was chosen to be the difference in chemical potential (free or available energy), instead of the gradient in simple concentration first proposed by Fick. Do we have any guarantee that the free energy of the mouse population will exceed that of the foxes?

To investigate this question we must either measure or estimate the respective entropies of mouse and fox. What, then, is the entropy of a fox? The fox itself is an open, highly nonequilibrium system. We could, of course, treat the fox as though it were a common substance, like salt or iron. Thus we would kill the fox, homogenize it, freeze-dry it, and look at how the heat capacity of the resultant material changes in the vicinity of absolute zero (−273.3°C). According to the established techniques of thermodynamics, this would give us a number to characterize the fox's entropy with respect to physical conditions at absolute zero. But what significance would this number have in relation to a living, metabolizing fox, much less to one that is free and actively feeding within its ecosystem?

The fact is that we really cannot *measure* the entropy of a fox in any meaningful way. If we do not know the value of the entropy, we cannot estimate the available energy, so all question about the actual direction of the free energy gradients for foxes vis-à-vis mice becomes pure speculation. The bitter truth is that we cannot quantify a force conjugate to the process of foxes feeding on mice. We might make some good guesses at what factors regulate this trophic process and build these factors into an ecological model, but this type of identification is far too case-specific to qualify as a proper thermodynamic force. In all but the simplest cases, thermodynamic forces remain obscure. The question is, do they exist at all?

Perhaps this question should be rephrased, "Do we really want thermodynamic forces to exist in general?" The processes that Onsager cites, such as thermal and mass diffusion, are spontaneous in nature. They are what happens whenever nothing else intervenes. They require *nothing* to elicit them. In this sense, spontaneous processes are like Newton's straight-line inertial motion. We recall that Newton defined force as that which intervenes to yield motion that deviates from straight-line translation. If the founders of irreversible thermodynamics had wished to follow Newton's thinking, they should have reserved the identification of forces for those agencies that invoke *nonspontaneous* behaviors, rather than the opposite. In their apparent zeal to build a description of irreversible systems upon a Newtonian framework, the founding fathers seem to have imported the wrong model. They were fixated on the assumption that natural processes are not inherently changing.

To summarize, the legacy of thermodynamics consists of two phenomenological laws, one of which appears to have little utility in ecology. Furthermore, in their attempts to extend physical notions to living, nonequilibrium systems, thermodynamicists have postulated the existence of generalized forces that remain quite obscure in the ecological realm. If forces were to appear in ecosystems, they would do so in a guise quite opposite to that which Onsager has formulated. These failings notwithstanding, thermodynamics did provide a coherent phenomenological framework that inherently was more defensible than were the suppositions of the atomists. Accordingly, we wish to apply second law notions to the tasks at hand, which are (a) to build, in analogy with nonconservative concepts (such as free energy), a system property that quantifies the relational form of constituent ecosystem processes, and (b) to quantify the causality behind individual ecosystem transformations in a very general way, i.e., to identify a surrogate for "ecological force."

2.7   Quantum Physics: Uncertainty Anew

The advent of irreversible thermodynamics occurred at a significant time, just *after* the revolution in quantum physics. Onsager, a Danish chemical engineer, could hardly have been unaware of this radical new perspective on submolecular events that was coalescing in, of all places, his very own backyard (Copenhagen). His formulations make it clear, however, that he shared none of his physicist colleagues' growing discontent with the Newtonian view. He must have felt that indeterminacy on the microscopic scale had little to do with the origins of irreversibility on the macroscale, and he was not alone in this opinion. Most contemporary scientists probably hope as well that indeterminacy, that break with Newton, is confined entirely to the netherworld of minuscule distances and minute times set by the value of Planck's constant.

In my opinion no greater evidence exists for the consensus by which the Newtonian worldview was preferred over the messy indeterminacy of microscopic events than the overwhelming regularity with which the new discipline is referred to as quantum "mechanics." True enough, there are formal parallels between classical mechanics and quantum physics. But the assumptions underlying the latter are so radically different from those of the former that quantum physics should be considered a fundamentally new way of looking at nature, and not a "mechanics" at all (Dicke and Wittke 1960). The quantum scenario forced a change in worldview to a far greater extent than did the statistical mechanics of Maxwell and Boltzmann. Through the lens of statistical theory, it was at least possible that one might reconcile the determinism of the macroscopic world with the confusion that reigns in the molecular realm. It was unclear, however, just how to regard the nature of microscopic events. They could be seen as arbitrary, stochastic, and refractory to any attempt to discover regularities in them—in which case, the probabilities used to describe the motions of atoms would reflect the *ontological* and objective nature of events at small scales. Alternatively, the molecules of a gas could be thought of as behaving in strict Newtonian fashion, tracing out reversible pathways through the space in which they were confined. In principle, Laplace's demon could know the past and future of the system; our own inability to do so would simply be a problem in *epistemology*.

In any case, the results of statistical mechanics obtain regardless of whether we assume that probabilities reflect nature or just our ignorance of it. Not so, however, with quantum physics. The indeterminacy of quantum phenomena has practical as well as conceptual roots. From the experimentalist's viewpoint it is no longer possible to believe that one can measure both the position and the momentum of small particles with arbitrary accuracy, because Werner Heisenberg was able to quantify inviolable limits on measurement precision. These limits in turn are a consequence of the very nature of particulate matter. For quantum indeterminacy appears to arise from attempts to apply two complementary but strictly incompatible models to the description of the same physical entity—say, an electron (Moore 1962). The particle model is derived from the concept of indivisible units of matter; it is basically an atomistic theory. The wave model arises from the notion of a continuous and infinite field. If the wavelength or frequency of an electron is to take on a precise value, and thereby an exact energy and momentum, the wave must have infinite extent, i.e., a completely indeterminate position. Conversely, to confine a wave within a finite region of space requires that waves of different frequencies be superimposed, thereby degrading the precision with which the momentum can be estimated.

At very small scales one loses touch not only with practical considerations, such as position and momentum, but with the nature of reality as well. Subsequent research into phenomena at subatomic dimensions has revealed a world so bizarre that the very idea of Newtonian behavior at such dimensions borders on the absurd. Although it was thermodynamics that first breached the seemingly impregnable walls of Newtonianism, it was quantum physics that revealed a yawning chasm in its integrity.

2.8　The "Grand Synthesis": Containing Uncertainty

If Onsager and colleagues were aware of the ongoing quantum revolution and chose to ignore it, they were hardly alone. Ronald A. Fisher by all accounts was also quite familiar with the ideas unfolding in physics during the early years of this century—but as a mathematician he was captivated more by the elegant statistics that Maxwell and Boltzmann had used in creating statistical mechanics. Just as those nineteenth-century giants had reconciled the deterministic world of everyday experience with the apparently stochastic netherworld of atoms, so Fisher set out to employ the same mathematics to demonstrate how the gradualist views of Darwin and later biometricians were compatible with the discrete characteristics featured in Mendelian genetics. To accomplish this, he had to assume a large population of *randomly* breeding individuals. Characters of the organism or phenotype were assumed to be determined not by a single gene, but by many genes that are constantly changing in Mendelian fashion. Fisher tracked the trajectories of gene frequencies in the same probabilistic spirit with which Maxwell, Boltzmann, and Gibbs tracked arrays of gas molecules (Depew and Weber 1994). The culmination of Fisher's labors was his "fundamental theorem of natural selection" whereby the fitness (reproduction) of gene frequencies is said to be maximized in much the same way as entropy was maximized in Boltzmann's scenario. While Boltzmann's direction was toward increasing disorder, however, Fisher's gene profiles progress in the direction of ever-increasing constraints. In Fisher's view we inhabit a "two-tendency universe," where the tendencies progress in opposite directions (cf. Hodge 1992).

Thus, even though it was to Maxwell and Boltzmann rather than Planck and Heisenberg that Fisher looked for guidance, the unintentional result of his theory was to open Pandora's box sufficiently to allow chance to emerge from its confinement in the world of subatomic particles and to establish a beachhead in biological theory. Although his theory was deterministic in the larger sense that genes control evolutionary behavior, it certainly was not determinism from the bottom in the Newtonian sense of Laplace.

If indeterminacy was an irritant thrust into the comfortable shell of biology, still it seemed possible to Fisher to contain the offending element and turn it into a pearl. The central idea of the Darwinian tradition is that natural selection is a two-step process (Mayr 1978): variation arises independently of selection, which in turn shapes and gives variation a direction (Depew and Weber 1994). Chance might be associated with variation, but selection continues to do its work with the same Newtonian regularity, linearity, and gradualism that Darwin had assumed. Thus, biologists eventually became aware that chance is inevitable at the molecular scale—the scale where variation arises, and where all those other unruly events occur that physicists treat using quantum theory. But Fisher had clearly marked the confines of chance in biology.

By this path we arrive at the current neo-Darwinian view of how change appears during the course of evolution. It is a strictly bipartite account of causation, which I have chosen to call "causality at the periphera." To follow the narrative requires that we constantly switch scales, alternating between the netherworld of Boltzmann, where variation originates, and the regularities of Newton, where the environment exerts its selection pressure. Put another way, causes may originate at both ends of the spectrum—in the subatomic domain, or somewhere out in the celestial realm, where the Newtonian perspective is most appropriate. Once a cause appears at either level, however, it propagates toward the middle ground of organisms and everyday experience in conservative, Newtonian fashion. In this scenario causes are proscribed from arising within the field of immediate experience, for that would wreak too much havoc for the conventional wisdom to contain.

About variation at small scales, Fisher (1930) concluded that the reliability of the physical material was found to flow, not necessarily from the reliability of its ultimate components, but simply from the fact that these components are *very numerous* and *largely independent*. At the other end of the hierarchy of events, Newton showed us how the reliability of physical events is a consequence of a *few* components that are *rigidly* linked by laws. Ecology, and many of the social sciences, usually treat an *intermediate* number of components that are *incompletely linked*.

It is not at all clear that neo-Darwinian logic can be successfully applied to ecology—at least to date, no success seems to clamor for recognition. To the contrary, a few biologists, such as Brenner and Stent, already are on record with their negative prognosis for the interpolation of neo-Darwinism into developmental biology and ecology. As we have seen, those who look to the established body of thermodynamic theory for direction find a phenomenology that becomes progressively more inchoate and flawed the closer it approaches the realm of ecology. In fact, there appear to be no irrefutable claims by any discipline to encompass ecological phenomena. Theoretical ecologists are thus free to strike out and boldly paint their own picture of the natural world, and then to invite others in to have a look.

3

THE EMERGENCE OF ORDER

Having just finished the last chapter, a pessimist or cynic might remark that this treatise is devolving quite naturally toward a dead end. To a strict determinist, the idea of an open universe evokes an emotion akin to that experienced when staring into a yawning abyss. The reductionist might concede that probabilities are useful for quantifying confusing situations, but would argue that behind any indeterministic facade lies hidden a deterministic reality—whence Einstein's famous quote, "God does not play dice!" Why question, after all, a worldview that has achieved such enormous success in advancing the material welfare of humanity? There exists a legion of humanists and social philosophers who are better qualified to respond to this question than I. Still, a degree of deconstruction or backtracking was necessary before forward progress could be resumed. Otherwise, we might carry the deterministic model into areas where it diverges ever further from perceived reality, amplifying our chances for a bad fall. I now wish to shift gears and turn my attention toward identifying and connecting those elements that eventually will lead to a quantitative description of some unsung origins of order in living systems.

3.1   Popperian Post-Positivism

Suppose we ask, Is it possible to abandon universal determinism and still remain optimistic about this endeavor called science? Some notable philosophers think so. One was the late Karl R. Popper, whom (in my experience) many immediately associate with the conservative movement in the philosophy of science called logical positivism. Such connection notwithstanding, Popper's later writings reveal attitudes that hardly any philosopher of science would label conservative: he bids us abandon the notion of a causally closed universe (Popper 1982). To be sure, he says, it is always possible to cite phenomena in relative isolation as examples of strict determinism. But among the enormous welter of events that make up our world, the strictly mechanical comprise but a minor fraction. Of course, mechanisms are useful as ideal limits to which other phenomena conform to greater or lesser degrees. The universe in general, however, is *open*. In accounting for the reasons why some particular event happens, it is often not possible to identify all the causes, even if we include all levels of explanation: there will always remain a small (sometimes infinitesimal) open window that no cause covers. This openness is what drives evolution. It is only by acknowledging such lacunae that we embark upon the pathway to a solid "evolutionary theory of knowledge."

These are revolutionary ideas. Popper not only paints Newtonian determinism as generally inadequate, but indicts more complicated frameworks as well. The neo-Darwinian scenario, or what I have called "causality at the peripheries," is itself called into question. Indeterminacy does not originate just in the netherworld of molecular confusion and then propagate to larger scales, driven as though ratcheted up by natural selection—it can arise at *any* level. Popper's idea of an open world challenges hierarchialist and Aristotelian theories as much as it does reductionistic, Newtonian ones. Aristotelians would agree that reductionism does not account for all reality, but for them that window could be closed by searching the focal and higher levels for sufficient causes. Popper would rejoin that even the full causal spectrum invariably has holes in it.

At first glance, Popper's emphasis on causal incompleteness seems to render mysterious what our intuition perceives as a degree of order and regularity in the world. If he is right, why don't things fly apart? Are we to abandon the concept of causality behind order as Peters (1993) recommends we do in ecology, or Tolstoy (1911) in history? Or should we surrender to the nominalists and absurdists, who deem all order and organization illusory?

Fortunately, neither of these extremes is necessary. Popper's world, though open, is not wholly without form. Those agencies that keep reality from dissolving into total randomness Popper (1990) calls "propensities." In his opinion, we inhabit a "world of propensities." They are the loose glue that keeps the world from flying apart. Propensities are the tendencies that certain processes or events *might* occur within a given context (Ulanowicz 1996). The subjective "might" connotes a probabilistic aspect to propensities. In fact, Popper's initial example could have come from any textbook on probabilities: Suppose we estimate the probability that a certain individual will survive until twenty years from the present, say to a particular day in 2017. Given the age, health, and occupation of that individual, we may use statistics on the survival of past similar individuals to estimate the probability that our subject will survive until 2017. As the years pass, however, the probability of survival until the given date does not remain constant: it may increase if the person remains in good health, decrease if accident or sickness intervenes, or even fall irreversibly to zero in the event of death. What Popper wishes to convey with this simple example is that there is no such thing as an absolute probability. All probabilities are contingent to a greater or lesser extent upon circumstances and interfering events (Ulanowicz and Wolff 1991). While this is manifestly clear in the example just mentioned, it is mostly ignored in classical physics, which deals largely with events that are nearly isolated and not progressive in nature. In classical physics events are either independent or rigidly coupled in mechanical fashion—if *A* occurs, then *B* follows in lockstep fashion: *B* is *forced* to follow *A*.

What in physics, then, are called "forces," Popper regards as the propensities of events in near-isolation. The classical example is the mutual attraction of two heavenly masses for each other. The virtual absence of interfering events in this case allows very precise and accurate predictions. With only a well-defined force at play, the probability of a given effect subsequent to its eliciting force approaches unity. *Propensities in the limit of no interfering agencies degenerate (in the mathematical sense of the word) into forces.*

Propensities are those agencies that populate the causal realm between the "all" of Newtonian forces and the "nothing" of stochastic infinitesima. They can appear spontaneously at *any* level of observation because of interferences among processes occurring at that level. This circumstance highlights Popper's second difference between propensities and common probabilities: Propensities are not properties of an object; rather, they are *inherent in a situation*. Propensities always exist among, and are mutually defined by, other propensities. There are no isolated propensities in nature, only isolated forces or nothingness.

While it may be permissible to talk about the force of attraction between two heavenly bodies in a context that is almost vacuous, Popper maintains that the same reasoning cannot be applied to the fall of an apple from a tree: "Real apples are emphatically not Newtonian apples!" (1990:24). When an apple will fall depends not only upon its Newtonian weight, but also upon the blowing wind, biochemical events that weaken the stem, and so forth. Exactly what happens and when it happens is conditional upon any number of other events. For this reason, Popper says, "We need a calculus of relative or conditional probabilities as opposed to a calculus of absolute probabilities" (1990:16).

3.2   A Calculus of Conditional Probabilities

Conditional probabilities already have been encountered by most readers. In a system consisting of multiple processes, we might identify a suite of potential "causes": call them $a_1, a_2, a_3, \ldots, a_m$. Similarly, we may cite a list of observable "effects": say, $b_1, b_2, b_3, \ldots, b_n$. We could then study the system in brute empirical fashion and create a matrix of frequencies that contains as the entry in row i and column j the number of times (events) that a_i is followed immediately by b_j. The a_i and b_j can be defined to serve empirical ends. For example, a_i might signify the ith type of human behavior among a list of m activities (e.g., eating raw meat, running, smoking, drinking alcohol), and the b_j could denote the jth type among n kinds of cancer (such as that of the lung, stomach, pancreas, etc.). The resulting events matrix would be filled with the numbers of individuals who habitually participated in the ith behavior and thereafter exhibited the jth form of cancer. For our applications, a_i and b_j will be defined so as to portray the kinetic form of a system. We will find it useful, for example, to let a_i represent the event, "A small amount of material leaves compartment i." The b_j in their turn will signify, "A quantum of material enters compartment j." The ensuing events matrix will record the magnitudes of transfers of material from compartment i into compartment j.

A matrix showing the hypothetical number of times that each of four causes was followed by one of five outcomes is given in table 3.1. For the sake of convenience, exactly 1,000 events were tabulated. This allows us to use the frequencies of joint occurrence to estimate the *joint probabilities* of occurrence simply by moving the decimal point one digit to the left. For example, of all the events that occurred, in 19.3% of the cases cause a_1 was followed by effect b_2, and in 2% of the cases a_4 was followed by b_4, etc. We denote these joint probabilities as $p(a_i, b_j)$. Listed in the sixth column are the sums of the respective rows: thus, a_1 was observed a total of 269 times; a_2, 227 times; etc. Similarly, the entries in the fifth row contain the sums of their respective columns: effect b_4 was observed 176 times; b_5, 237 times; etc. These marginal sums are estimators of the *marginal probabilities* of each cause and

TABLE 3.1

Hypothetical Frequencies of Occurrence of Each of Five "Effects" (b)
Following Each of Four "Causes" (a)

	b_1	b_2	b_3	b_4	b_5	Sum
a_1	40	193	16	11	9	269
a_2	18	7	0	27	175	227
a_3	104	0	38	118	3	263
a_4	4	6	161	20	50	241
Sum	166	206	215	176	237	1000

effect. Thus, 26.3% of the times a cause was observed, it was a_3; or, 17.6% of the observed effects were b_4. We denote the marginal probabilities by $p(a_i)$ or $p(b_j)$.

A *conditional probability* is the answer to the question, "What is the probability of outcome b_j given that 'cause' a_i has just occurred?" The answer is easy to calculate. For example, if $i = 2$ and $j = 5$, then a_2 occurred a total of 227 times, and in 175 of those instances the result was b_5; therefore, the conditional probability of b_5 occurring, given that a_2 has happened, is estimated by the quotient 175/227, or 77%. In like manner, the conditional probability that b_4 happens, given that a_1 has just transpired, is 4.1%; etc. If we represent the conditional probability that b_j happens given that a_i has occurred by $p(b_j|a_i)$, then we obtain the general formula developed in the eighteenth century by Thomas Bayes:

$$p(b_j|a_i) = p(a_i, b_j)/p(a_i).$$

It is well to pause at this point and consider what sort of system might have given rise to the frequencies in table 3.1. Unless the observer was unusually inept at identifying the a_i's and b_j's, then the system did not behave in strictly deterministic, mechanical fashion. We see that most of the time a_1 gives rise to b_2, a_2 to b_5, and a_4 to b_3. But there is also a lot of what Popper calls "interference"—situations like those in which a_4 yielded b_1, which were occasioned either by some external agency or by the interplay of processes within the system. We also notice that there is significant ambiguity as to whether the outcome of a_3 will be b_1 or b_4.

If it were possible to isolate individual processes and study them in laboratory-like situations, then something like a mechanical description of the system might ensue. For example, we might discover that if we take great care to isolate processes, a_1 always yields b_2, a_2 gives b_5, a_3 invariably results in b_1, and a_4 in b_3. The same frequency counts taken

	b_1	b_2	b_3	b_4	b_5	Sum
a_1	0	269	0	0	0	269
a_2	0	0	0	0	227	227
a_3	263	0	0	0	0	263
a_4	0	0	241	0	0	241
Sum	263	269	241	0	227	1000

TABLE 3.2
Frequency Table Like Table 3.1, Except That Causes Were Isolated from One Another

from a collection of isolated processes could look something like those in table 3.2. Knowing a_i immediately reveals the outcome b_j in mechanical, lockstep fashion. As Popper noted, the conditional probabilities of force-effect pairs are all unity, or certainty. What is also interesting in table 3.2 is that b_4 is never the outcome of an isolated cause; it is therefore likely that in the natural ensemble, b_4 is purely the result of interaction phenomena.

We see that *whenever systems become "hardened," or more mechanical in nature, the propensities of events approximate forces, which are manifested by conditional probabilities of unity*. On the other hand, conditions typified by a system that gives rise to an events matrix like that in table 3.1 are obviously more confused or less orderly than those characterized by table 3.2. The process of going from a less orderly to a more organized pattern of behavior we will refer to under the rubric of "development." *During the course of development, propensities tend to become progressively isolated from their environment and in the end to approximate forces that are characteristic of mechanical behavior.* If Popper is correct, however, real systems never reach such a closed endpoint; they always remain embedded in the real, open world.

3.3   Formal and Final Agencies

What sort of agency could engender such development? Recall the remarks by Brenner and Stent: "We have to try to discover the principles of organization, how lots of things are put together in the same place" (Lewin 1984:1327). Therefore, we begin our search for ordering agencies by considering what sort of interactions might ensue when two processes occur in proximity to one another. There are three qualitative effects the first process could have on the second: it could be beneficial

(+); it could be detrimental (–); or it could have no effect whatsoever (0). The second process, in its turn, could have any of these same effects on the first. Hence, there are nine pairs of possible interactions: (+ –), (– 0), (– –), etc. One of these combinations, however, can give rise to behavior that is qualitatively very different from the other eight possibilities (cf. Bateson 1979). Furthermore, this particular relationship has the potential for generating decidedly nonmechanical behaviors.

Mutualism is defined as (+ +). It is a special case of positive feedback (DeAngelis, Post, and Travis 1986). Positive feedback can arise according to any number of scenarios, some of which involve negative interactions. (Two negative interactions taken serially can yield a positive overall effect.) "Mutualism" we shall take as "positive feedback comprised wholly of positive component interactions." Mutualism so defined need not involve only two processes; when more than two elements are involved, it becomes "indirect mutualism." A schematic of indirect mutualism among three processes or members is presented in figure 3.1. The plus sign near the end of the arrow from *A* to *B* indicates that an increase in the rate of process *A* has a strong propensity to increase the rate of *B*. Likewise, growth in process *B* tends to augment that of *C*, which in its turn reflects positively back upon process *A*. In this sense, the behavior of the loop is said to be "autocatalytic," a term borrowed from chemistry that means "self-enhancing." In an autocatalytic system, *an increase in the activity of any participant will tend to increase the activities of all the others as well.*

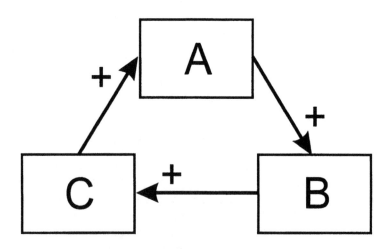

FIGURE 3.1.
Schematic of a hypothetical three-component autocatalytic cycle.

In keeping with Popper's idea of an open universe, we do not require that *A*, *B*, and *C* be linked together in lockstep fashion. To achieve autocatalysis we require only that the propensities for positive influence be stronger than cumulative decremental interferences. There is also an issue of the phasing of the influences. It is conceivable that the timing of sequential positive effects could result in overall negative feedback; such configurations are simply excluded from our definition of autocatalysis.

Many examples of indirect mutualism in ecology are subtle and require much elaboration (e.g., Bertness and Hacker 1994, or Bertness and Callaway 1994), but I have proposed one that is somewhat more straightforward (Ulanowicz 1995d). Found in freshwater lakes over much of the world, and especially in subtropical, nutrient-poor lakes and wetlands, are various species of aquatic vascular plants belonging to the genus *Utricularia*, or the bladderwort family (Bosserman 1979). Although these plants are sometimes anchored to lake bottoms, they do not possess feeder roots that draw nutrients from the sediments; rather, they absorb their sustenance directly from the surrounding water (figure 3.2). Let us identify the growth of the filamentous stems and leaves of *Utricularia* into the water column with process *A* in figure 3.1.

Upon the leaves of the bladderworts invariably grows a film of bacteria, diatoms, and blue-green algae that collectively is known as periphyton. There is evidence that some species of *Utricularia* secrete mucous polysaccharides (complex sugars) to bind algae to the leaf surface and attract bacteria (Wallace 1978). Bladderworts are never found in the wild without their accoutrement of periphyton. Apparently, the only way to raise *Utricularia* without its film of algae is to grow its seeds in a sterile medium (Bosserman 1979). Suppose we identify process *B* with the growth of the periphyton community. It is clear, then, that bladderworts provide an areal substrate that the periphyton species (not being well adapted to growing in the pelagic, or free-floating, mode) need in order to grow. Some species may even provide other subsidies to the periphyton film.

Component *C* enters in the form of a community of small, almost microscopic (ca. 0.1 mm) motile animals, collectively known as "zooplankton," which feed on the periphyton film. These zooplankton can be from any number of genera of cladocerans (water fleas), copepods (other microcrustacea), rotifers, and ciliates (multicelled animals with hairlike cilia used in feeding). In the process of feeding on the periphyton film, these small animals occasionally bump into hairs attached to one end of the small bladders, or utrica, that give the bladderwort its family name. When moved, these trigger hairs open a hole in the end of the bladder,

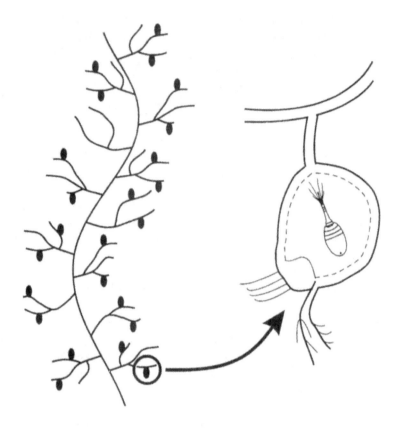

FIGURE 3.2.
Sketch of a typical "leaf" of *Utricularia floridana*, with detail of the interior of
a utricle containing a captured invertebrate.

the inside of which is maintained by the plant at negative osmotic pressure with respect to the surrounding water. The result is that the animal is sucked into the bladder, and the opening quickly closes behind it. Although the animal is not digested inside the bladder, it does decompose, releasing nutrients that can be absorbed by the surrounding bladder walls. The cycle of figure 3.1 is now complete (see figure 3.3).

Because the example of indirect mutualism provided by *Utricularia* is so colorful, it becomes too easy to get lost in the seemingly mechanical details of how it, or any other example of mutualism, operates. It is important, accordingly, to note that in any biological system the components maintain some plasticity or indeterminacy. Such is obviously the case with the periphyton and zooplankton communities, for their com-

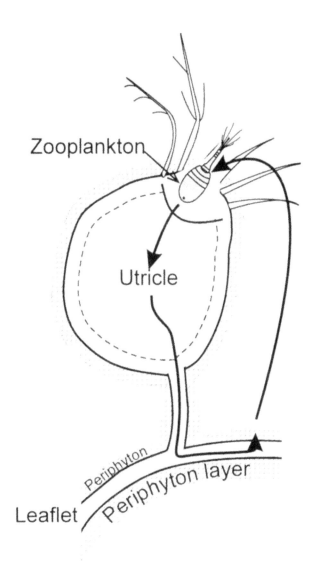

FIGURE 3.3.
Schematic of the autocatalytic loop in the *Utricularia* system. The macrophyte
provides a necessary surface upon which the periphyton (*striped area*) can
grow. Zooplankton consumes the periphyton, and is itself trapped in the
bladder and absorbed by the *Utricularia*.

positions change with various habitats. Plasticity applies as well over the longer time scale to *Utricularia* itself, which has evolved into numerous species, and even exhibits a degree of polymorphism over rather short intervals (Knight and Frost 1991). Such plasticity or adaptability contrasts with the usual situation in chemistry, where the reactants in any autocatalytic process are fixed, thereby contributing to the stereotypical image of autocatalysis as a "mechanism."

Although autocatalysis as mechanism may well pertain to most *chemical* examples, such identification is wholly inappropriate as soon as the elements that constitute the autocatalytic loop become adaptable. In general, autocatalysis is *not* a mechanism! Autocatalytic systems exhibit properties that, taken in concert, transcend the much-overused metaphor of nature-as-machine (Ulanowicz 1989).

As a first example, autocatalytic configurations, by definition, are *growth enhancing*. An increment in the activity of any member engenders greater activity in all other elements: the feedback configuration results in an increase in the aggregate activity of all members engaged in autocatalysis over what it would be if the compartments were decoupled. Of course, even conventional wisdom acknowledges the growth-enhancing characteristic of autocatalysis. Far less attention is paid, however, to the *selection pressure* that the overall autocatalytic form exerts upon its components. For example, if a random change should occur in the behavior of one member that either (a) makes it more sensitive to catalysis by the preceding element or (b) accelerates its catalytic influence upon the next compartment, then the effects of such alteration will return to the starting compartment as a reinforcement of the new behavior. The opposite is also true: should a change in the behavior of an element either make it less sensitive to catalysis by its instigator or diminish the effect it has upon the next in line, then even less stimulus will be returned via the loop.

Unlike Newtonian forces, which always act in equal and opposite directions, the selection pressure associated with autocatalysis is inherently *asymmetric*. Autocatalytic configurations impart a definite sense (direction) to the behaviors of systems in which they appear. They tend to ratchet all participants toward ever greater levels of performance.

Perhaps the most intriguing of all attributes of autocatalytic systems is the way they affect transfers of material and energy between their components and the rest of the world. Figure 3.1 does not portray such exchanges, which generally include the import of substances with higher exergy (available energy) and the export of degraded compounds and heat. The degradation of exergy is a spontaneous process mandated by the second law of thermodynamics—but it would be a mistake to assume that the autocatalytic loop is itself passive and merely driven by the gradient in exergy. Suppose, for example, that some arbitrary change happens to increase the rate at which materials and exergy are brought into a particular compartment. This event would enhance the ability of that compartment to catalyze the downstream component, and the change eventually would be rewarded. Conversely, any change decreasing the intake of exergy by a participant would ratchet down activity throughout the loop. The same argument applies to every member of the loop, so that the overall effect is one of *centripetality*, to use a term coined by Sir Isaac Newton: the autocatalytic assemblage behaves as a focus upon which converge increasing amounts of exergy and material that the system draws unto itself (figure 3.4).

Taken as a unit, the autocatalytic cycle is not acting simply at the behest of its environment: it actively creates its own domain of influ-

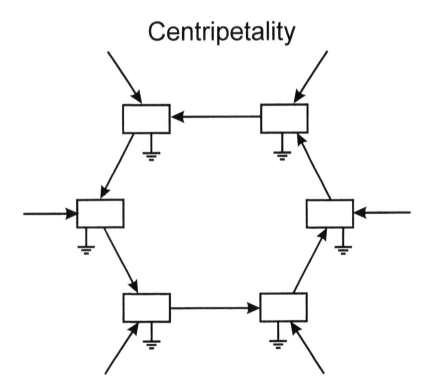

Centripetality

FIGURE 3.4.

Autocatalytic cycle exhibiting centripetality. The loop develops so as to draw into itself various resources necessary for its functioning.

ence. Such creative behavior imparts a separate identity and ontological status to the configuration, above and beyond the passive elements that surround it. We see in centripetality the most primitive hint of entification, selfhood, and id. In the direction toward which the asymmetry of autocatalysis points we see a suggestion of a telos, an intimation of final cause (Rosen 1991). Popper put it all most delightfully: "Heraclitus was right: We are not things, but flames. Or a little more prosaically, we are, like all cells, processes of metabolism; nets of chemical pathways" (1990:43).

To be sure, autocatalytic systems are contingent upon their material constituents and usually also depend at any given instant upon a complement of embodied mechanisms. But such contingency is not, as strict reductionists would have us believe, entirely a one-way street. By its very nature autocatalysis is prone to *induce competition*, and not merely among different properties of components (as discussed above under selection pressure): its very material and (where applicable) mechanical constituents are themselves prone to replacement by the active agency of the larger system. For example, suppose that A, B, and C are three sequential elements comprising an autocatalytic loop as in figure 3.5a, and that some new element D appears by happenstance, is more sensitive to catalysis by A, and provides greater enhancement to the activity of C than does B (figure 3.5b). Then D either will grow to overshadow B's role in the loop, or will displace it altogether (figure 3.5c).

In like manner one can argue that C could be replaced by some other component E (figure 3.5d), and A by F, so that the final configuration D-E-F contains none of the original elements (figure 3.5e). (Simple induction will extend this argument to an autocatalytic loop of n members.) It is important to notice in this case that the characteristic time (duration) of the larger autocatalytic form is longer than that of its constituents. The persistence of active form beyond present makeup is hardly an unusual phenomenon—one sees it in the survival of corporate bodies beyond the tenure of individual executives or workers, or in plays, like those of Shakespeare, that endure beyond the lifetimes of individual actors. But it also is at work in organisms as well. Our own bodies are composed of cells that (with the exception of neurons) did not exist seven years ago, and the residencies of most chemical constituencies, even those comprising the neural synapses by which are recorded long-term memory in the brain, are of even shorter duration; yet most of us still would be recognized by friends we have not met in the past ten years.

Overall kinetic form is, as Aristotle believed, a causal factor. Its influence is exerted not only during evolutionary change, but also during the

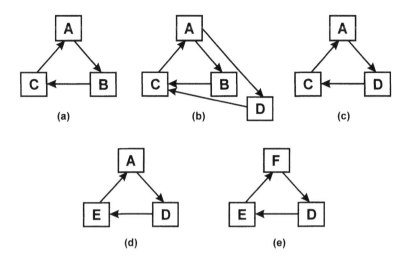

FIGURE 3.5.
Successive replacement of the components in an autocatalytic loop:
(a) The original loop, as in figure 3.1. (b) Component D appears and is more
effective in the feedback loop. (c) D eventually replaces B. (d) C is likewise
replaced by E. (e) Finally, A is replaced by F, so that none of the original
components remain.

normal replacement of parts. For example, if one element of the loop
should happen to disappear, for whatever reason, it is (to use Popper's
own words) "always the existing structure of the . . . pathways that determines what new variations or accretions are possible" to replace the
missing member (Popper 1990:44).

The appearance of centripetality and the persistence of form beyond
constituents make it difficult to maintain hope for a strictly reductionist,
analytical approach to describing organic systems. Although the system
requires material and mechanical elements, it is evident that some
behaviors, especially those on a longer time scale, are, to a degree,
autonomous of lower-level events (Allen and Starr 1982). Attempts to
predict the course of an autocatalytic configuration by ontological
reduction to material constituents and mechanical operation are, accordingly, doomed over the long run to failure.

It is important to note that the autonomy of a system may not be
apparent at all scales. If the observer's field of view does not include all
the members of an autocatalytic loop, the system will appear linear in
nature, and it will seem that an initial cause and a final result can be iden-

tified (see figure 3.6). The subsystem can appear wholly mechanical in its behavior. For example, the phycologist who concentrates on identifying the genera of periphyton found on *Utricularia* leaves would be unlikely to discover the unusual feedback dynamics inherent in this community. Once the observer expands the scale of observation enough to encompass all members of the loop, however, then autocatalytic behavior with its attendant centripetality, persistence, and autonomy *emerges* as a consequence of this wider vision.

The emergence of qualitatively different phenomena carries different epistemological overtones in post-Newtonian science. At the time of Laplace there was great optimism that we could grasp reality, and that reality was decidedly mechanical in essence. Anything else had to be illusory, because it always could be disassembled into fundamental parts that operated in a fixed, lawful manner. Thus arose the vision that human attributes (such as free will, intention, and consciousness), and even the integrity of nonsentient living organisms, do not possess the same ontological status as the hard-and-fast interactions of elementary particles. The "higher" forms could only be epiphenomena—somewhat like the images of actors when a film or videotape is played.

The past century, however, has taught us a little more humility concerning our ability to perceive reality directly. There always remains a veil that separates our models of the world from reality itself (Allen and Hoekstra 1992). Universality, whereby laws apply uniformly over all scales of time and space, is now viewed in many quarters with much skepticism. Our new picture of the world is more "granular" in texture. Laws are now thought to apply to only a finite range of spatial and temporal scales. As one moves beyond those limits, the effects of events at the starting scale gradually lose their meaning in the new context, while at the same time qualitatively new phenomena emerge. Attempts to stretch explanations and applications of certain terms over several hierarchical levels should be scrutinized carefully. For example, any effort to trace volition in humans to roots in quantum physics is likely to be regarded as comic (Depew and Weber 1994). Similarly, an attempt to trace all social behavior to genetic origins is likely to be an ill-considered endeavor (Wilson 1975).

A conservative systems theory has evolved that uses a triadic approach to explanation (Salthe 1985). First, the system is carefully defined at a particular level of resolution, called the focal level. The perspective is called triadic because attention is paid to events that occur at three different levels—the focal level, the next-higher level, and the one immediately below the focal (figure 3.7). Heretofore, virtually all explanations

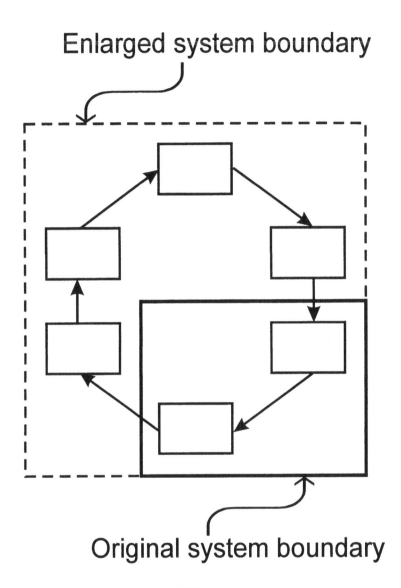

FIGURE 3.6.

Two hierarchical views of an autocatalytic loop. The original perspective
(*solid line*) includes only part of the loop, which therefore appears to function
quite mechanically. A broader vision encompasses the entire loop, and with it
several nonmechanical attributes.

consisted of relating happenings at the focal level to their mechanical or efficient causes at the next-lower level. Occasionally, but far less frequently, someone might pay attention to constraints that the larger and slower-changing phenomena at the next-higher level placed on system behavior. Almost no time, however, was spent searching for explanations at the focal level. The idea that causes might be originating *at* the focal level was simply out of the question, precisely because that level is assumed to be created by its relations with the other two.

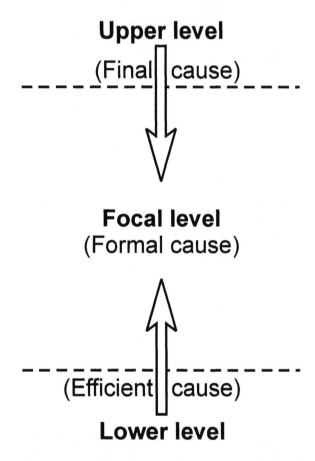

Upper level
(Final cause)

Focal level
(Formal cause)

(Efficient cause)
Lower level

FIGURE 3.7.

The triadic view of causalities as they affect a system. Formal cause arises at the focal level, whereas efficient cause operates from below. Final causes are impressed from above.

In our consideration of autocatalytic systems, however, we have seen that agency can arise quite naturally at the very level of observation. This occurs via the relational form that processes bear to one another. That is, autocatalysis takes on the guise of a *formal* cause, *sensu* Aristotle. Nor should we ignore the directionality inherent in autocatalytic systems by virtue of their asymmetric nature: such rudimentary telos is a very local manifestation of final cause that potentially can interact with similar agencies arising in other parts of the system.

To recapitulate, this examination of indirect mutualism has revealed that autocatalytic systems can possess at least eight properties. *Autocatalysis induces (1) growth and (2) selection. It exhibits (3) an asymmetry that can give rise to the (4) centripetal amassing of material and available energy. The presence of more than a single autocatalytic pathway in a system presents the potential for (5) competition. Autocatalytic behavior is (6) autonomous, to a degree, of its microscopic constitution. It (7) emerges whenever the scale of observation becomes large enough, usually in the guise of an Aristotelian (8) formal cause.*

To connect this description with those normally given in thermodynamics, we note that thermodynamic variables are usually distinguished as either extensive or intensive. An *extensive* property depends upon the size of the system: if a homogeneous system is divided in half, each portion will contain one-half of the given extensive attribute, such as mass or energy. An *intensive* variable, on the other hand, does not depend on the size of the system; for example, the pressure in both halves of a separated system should remain unaffected by the partitioning. In the case of autocatalysis, however, we see that this single agency can affect both extensive and intensive system properties. That is, we identify growth and centripetality as extensive attributes that tend to inflate overall activity. On the other hand, selection and competition serve to prune those components and linkages that are less efficient at participating in the autocatalytic scene. The overall effect on a network is illustrated schematically in figure 3.8, where the dots represent the system components and each arrow symbolizes a process by which one component influences another. Figure 3.8a represents how interactions might appear in an inchoate system. As autocatalysis develops in the system, the overall level of activity increases (represented by the thickness of the arrows), and those processes most active in the feedback dynamics grow at the expense of nonparticipating activities; eventually, the system comes to resemble more the one depicted in figure 3.8b.

One could easily create events frequency tables that correspond to these two system configurations. The arrows might represent, for exam-

ple, the frequencies of feeding events in an ecosystem. The ecosystem pictured in figure 3.8a would yield a frequency matrix with diffuse entries, as in table 3.1; the configuration in 3.8b would yield frequencies of feeding events concentrated in fewer matrix cells, as in table 3.2. We recall that the transition from table 3.1 to table 3.2 is indicative of the process of development. Thus, we may conclude that indirect mutualism, or autocatalysis, can act as a formal agency that imparts development to living systems.

We now see that, even in an open world, systems do not only fall apart and disintegrate: given enough usable energy and materials, configurations of processes can arise (spontaneously or otherwise) that maintain and even create order at the focal level. Such creation of order in no way violates the second law of thermodynamics. As Prigogine (1967) pointed out, the second law applies to the universe as a whole, not to a specific part of it. A system may become progressively more ordered, so long as conditions outside the system *at other hierarchical levels* become less determinate at an even faster rate.

4

QUANTIFYING GROWTH AND DEVELOPMENT

4.1   A Macroscopic Image of Ecosystems

It is most reassuring to discover that the causal holes in the fabric of nature do not necessarily render the world tatters. Formal and final agencies may arise at any hierarchical level to impart order and cohesion to any system. Of course, larger entities remain contingent upon material configurations of constituent processes at smaller scales. Thus, when Jeffrey Wicken (1984:108) writes that "life is based on informed kinetic stabilizations requiring chemical flows," much of the implied information resides in the material genomes of the elements that constitute the pattern of exchanges. We have noted, however, that in an open universe, any configuration at a given scale can be somewhat autonomous of what transpires at finer resolutions. It becomes permissible, therefore, to concentrate on description at the focal level, while keeping implicit most of the contributions from finer scales. We are free, for example, to consider the growth and development of ecosystems without explicitly mentioning genes or the DNA embedded in them. We can even conceive of a totally self-consistent and coherent body of phenomenological observations that explicitly mentions only agencies at the focal level.

To some readers, talking about ecology without mentioning genes may sound like heresy, but such an approach is hardly without precedent in other fields. There exists a school of thermodynamicists, for example, that insists upon the sufficiency of macroscopic narration (which is not intended to constitute full explanation). As a student in chemical engineering science, I was made to toe this party line. If, in response to any question on thermodynamics, a student should utter or write the words "atom" or "molecule," the answer was summarily judged incorrect. Thermodynamics, according to this dogma, was a self-consistent body of phenomenological observation quite divorced from any theory of atoms. As T. F. H. Allen has suggested, "A paradigm is a tacit agreement *not* to ask certain questions" (Ahl and Allen forthcoming). It was only the atomic hypothesis, after all, that had been at risk during the conflict between thermodynamics and Newtonianism, not the body of macroscopic phenomenology.

In that same vein, we press forward here to consider the development of ecological systems. It is not that ecologists would make light of the startling discoveries and advances recently made in the field of molecular biology. Everyone recognizes the contributions of "microscopic" biology to our understanding of the natural world and, if applied with intelligence and some humility, to our own welfare. It is simply that, as Brenner implied, these things do not tell us all that we want to know about ecosystems. We wish to concentrate on the phenomena that are proper to *ecological* description. As ecologists we are entitled to choose our own focal level, and free to search for causes *appropriate to* that level.

Just what are the phenomena that describe an ecosystem? I wish to emphasize first the transformations of materials and energy that accompany processes such as feeding, decay, excretion, etc. For example, in an estuary such as Chesapeake Bay the striped bass (*Morone saxatilis*) feed upon bay anchovy (*Anchoa mitchelli*). Suppose we want to know how much of the prey population was consumed by the predator within a particular area over a given period of time. For reasons to be discussed later, we stipulate that this transfer be quantified in terms of some "currency" that is either material (e.g., amounts of carbon, nitrogen, or phosphorus) or energetic (e.g., kcal.) in kind. Thus, the feeding by striped bass upon anchovy in the middle of Chesapeake Bay has been reported as 17.9 mg carbon per square meter per year (Baird and Ulanowicz 1989). Alternatively, the same transfer could be quantified in terms of the amount of nitrogen (4.6 $mg/m^2/y$) or phosphorus (0.27 $mg/m^2/y$) exchanged (Baird, Ulanowicz, and Boynton 1995).

Under the physical conditions that surround ecosystems, material and energy are conserved in any transaction. Although conservation of medium is not strictly required for the analyses that follow, it is a bookkeeping device useful for inferring flows that are hard to measure from those that are more accessible. Conservation also provides a way of estimating whether we have considered all possible transfers. For example, we might begin by conceiving of the ensemble of living populations in our ecosystem as a collection of boxes or nodes that are connected with each other via arrows that represent feeding linkages so as to form one of Popper's "nets of chemical pathways." One of the first things we discover in trying to balance material into and out of any compartment over a sufficiently long time is that much of the material that leaves the box does so in the form of nonliving organic matter (e.g., feces, urine, dead bodies, and other forms of detritus). Hence, to complete our description of the ecosystem it becomes necessary to identify the various exchanges of dead organic matter.

Even after nonliving forms of material have been incorporated into the system, however, it generally will be the case that exchanges among the identified components will not account for all transactions: the origins and fates of some material always lie outside the defined system. That is, each compartment of the system may, in principle, receive material or energy from outside the system. A plant captures energy from the sun. The immigration of motile species is another way to import material into a compartment. In aquatic and marine systems, the importation of nonliving materials along with currents flowing into the body of water is a common subsidy. Likewise, material and energy are always being exported out of the system. As regards energy, this can occur in two distinct qualitative ways. Some energy invariably leaves the system as heat—i.e., energy in its degraded form. With rare exceptions (for instance, the clumping of bees in winter) this energy cannot be used to create or maintain biological order; it has been dissipated. Other energy leaves the system embodied in organic or chemical constituents, and in those forms it may be imported by elements of some other system at the same hierarchical level. This is an export of useful energy.

As mentioned, material is strictly conserved within the ecosystem. Although some chemical forms cannot easily be reused by the ecosystem (carbon in graphite form, for example), all the chemical elements vital to living systems can in principle reenter the life process. Nevertheless, the convention in ecology has been to identify the energetically most degraded state of a vital element (e.g., carbon dioxide or molecular nitrogen, N_2) and exclude that particular form from the system. Any conversions into this base state are regarded as "dissipations." For example, respiration is the process by which organisms release carbon to the environment in the form of carbon dioxide. On most ecosystem carbon flow networks respirations appear in the guise of a dissipative export.

To summarize, there are four separate categories of flows that need be estimated in order to complete the inventory of material exchanges in an ecosystem (figure 4.1): (1) Internal transfers, which we shall refer to as

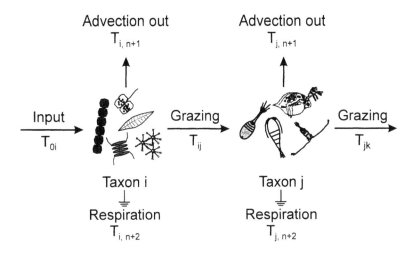

FIGURE 4.1.

The four types of exchanges used to quantify ecosystem flow networks:
(T_{ij}) Internal flow from arbitrary compartment i to any other taxon, j, of an
n-compartment system. (T_{0i}) External inputs to the arbitrary compartment i.
($T_{i, n + 1}$) Export of usable resources from unit i. ($T_{i, n + 2}$) Dissipation of
resources from system element i.

T_{ij}, where the subscripts mean the amount has been transferred from
compartment or species i to component j. (Compartments usually are
numbered 1, 2, 3, . . . , n, so that i and j can be any integer from 1 to n.)
(2) Imports from the external world into compartment i. (3) Exports of
usable material or energy from compartment i to another comparable
system. (4) Dissipation of energy or conversion of material by component i into its base form.

In any real ecosystem only a fraction of the conceivable transfers
actually takes place. For example, if compartment 1 represents a sea-
grass in a littoral (shallow water) marine community, and 6 is the desig-
nation for a shark, it is conceivable, but not likely, that the transfers T_{16}
or T_{61} could occur directly. In ecosystems that have been parsed to any
reasonable degree of resolution the percentage of possible direct connections that actually occurs is usually well under 20%. (Indirect connections, of course, are far more numerous [Patten and Higashi 1991].)

Figure 4.2 is a schematic of the carbon transfers that occur among
thirty-six principal components of the ecosystem inhabiting the meso-

haline (intermediately salty) reach of Chesapeake Bay (Baird and Ulanowicz 1989). The autotrophs, or plants, are represented by two bullet-shaped nodes (H. T. Odum 1971). The heterotrophic, or animal, components are depicted as hexagons; the nonliving "storages" of organic matter appear as "birdhouses." Respirations, or dissipations, take the form of "ground connections" in electrical circuitry.

Although on first sight figure 4.2 may seem terribly complicated or busy, in comparison to everything that is happening in the ecosystem it actually portrays only a very small fraction of events and behaviors. As a minimalistic representation of an ecosystem, it makes Newton's description of the physical universe seem absolutely baroque.

Representing ecosystems in terms of their underlying webs of trophic interactions seems to be an American tradition. It traces back to Raymond Lindeman (1942), a student of G. Evelyn Hutchinson. With some justification, such schematic portrayals have been characterized as "brutish" representations of the highly complicated, almost ethereal entity that is an ecosystem (Engelberg and Boyarsky 1979). Having inveighed against oversimplification, why suddenly resort to a picture of ecosystems that is both highly material and very minimalistic?

It would be easy for me to dissemble at this point and say that simplified representations, like that in figure 4.2, are an attempt to condescend to the positivist tradition in science—to demonstrate to skeptics in their own material terms a reality that transcends Newtonian beliefs. In point of fact, however, my own experience was quite the reverse. I did not start with the abstract and proceed to the material. Rather, I began to analyze this quite minimalist scheme in traditional, reductionist fashion and gradually came to realize how several seemingly unrelated regularities cited by others (E. P. Odum 1969) could be unified into a single mathematical statement (Ulanowicz 1980). Such an amazing coincidence begged for explanation, and nothing within the Newtonian or neo-Darwinian traditions appeared to suffice. Thus began my search for a larger context in which to place the story of ecosystem growth and development.

FIGURE 4.2.

Estimated flows of carbon (mg/m^2/y) among the thirty-six principal components of the mesohaline Chesapeake Bay ecosystem (Baird and Ulanowicz 1989). "Bullets" represent autotrophic system elements (plants); hexagons, heterotrophic taxa (fauna); and "birdhouses," nonliving storages. (*DOC* = dissolved organic carbon; *POC* = particulate organic carbon.) Numbers inside each box are the standing stocks in mg/m^2.

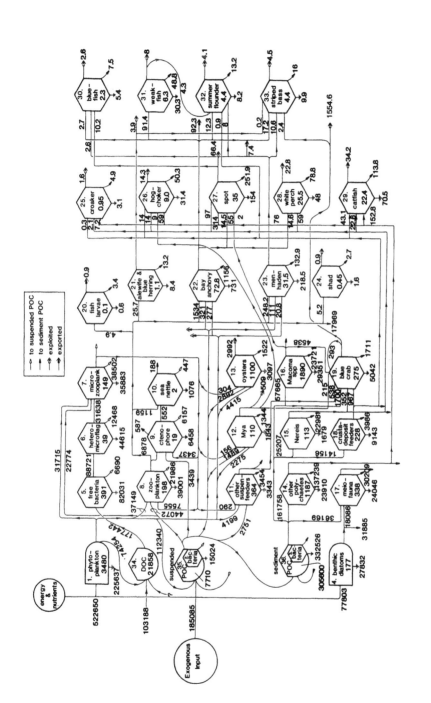

One must take seriously the criticism of Popper's propensities that a causal link must not be inferred from a probabilistic observation (as in, for example, the mistake of "Post hoc, ergo propter hoc"). This leads to the second reason for narrowing subsequent analyses to include only exchanges of a material currency: such a requirement guarantees a minimal causal linkage (that of material causality). We should be reluctant to proffer anything as science that cannot somehow be measured in the field or the laboratory. Hence, we pose questions like, "Given that a quantum of material is now embodied in population X, how does that fact constrain whether it will appear as part of population Y or Z during the next interval of time?" To pursue an answer we may set up an events table (like tables 3.1 and 3.2) and use it to estimate the appropriate probabilities. There may *or may not* be a mechanism that gives rise to the observed probabilities. If not, we could still weigh whether the probabilities might be the result of a formal or final agency. What we *cannot* do under any circumstances, however, is dismiss a propensity calculated on such a material basis as being devoid of causal content. The transfers palpably occurred; whenever they happen with significant frequency, the underlying propensity *demands* our serious consideration.

There are, of course, innumerable aspects of ecological systems that cannot be quantified as readily as trophic transfers. If, however, such attributes and events have any significance whatsoever, their effects must be impressed into the pattern of trophic flows. For example, the ethologist may be interested in mating behavior in birds that is cued by colors and characteristic movements that might resist quantification. Still, whether or not they lead to successful mating affects the size of the bird population, and thereby its aggregate demand for food and its availability as prey. The ethology leaves its mark upon the pattern of exchanges.

An example of how a nontransfer mechanism can influence the pattern of flows is to be found in the feedback dynamics of the *Utricularia* communities. The positive effects of periphyton upon zooplankton and of zooplankton upon *Utricularia* take the form of direct material transfers. The benefit of *Utricularia* to periphyton, on the other hand, does not: the macrophyte simply provides an areal substrate upon which the periphyton can grow. (As mentioned, there may be a mucilaginous subsidy provided by the plant to the colonizing periphyton, but such a donation is not a necessary element of the feedback cycle.) Remove that nonmaterial linkage (areal support) and the positive feedback dynamics cease, causing significant changes in the overall flow structure (Ulanowicz 1995d).

4.2   Systems Growth

Limiting the discussion of ecosystem networks to palpable material transfers is a step backward that allows more than compensating progress forward. Such weighted flow networks now make it possible to quantify the effects of indirect mutualism that were portrayed pictorially in figure 3.8. We recall that autocatalysis affects both the extensive and intensive attributes of the system in which it acts. Let us treat the extensive aspect first. We have seen how autocatalysis works to foster system growth. More precisely, it enhances overall system activity. Aggregate system activity is relatively easy to quantify. Each arrow in figure 4.2 represents an ecosystem process, and the magnitude associated with each arrow, the intensity of that process. By summing the magnitudes of all the arrows (for all four types of flow), we directly characterize the total system activity. All the transfers in this network add up to 4,116,200 mg C/m^2/y. In economics, this sum is called the "total system throughput" (TST), a term that has been imported into ecology (Hannon 1973; Finn 1976).

We most commonly associate the term *growth* with an increase in the extensive properties of a system. This could mean either that the system increases in spatial extent, or that its level of activity rises. In economics, emphasis is placed upon the latter, because it addresses better the relational aspects of communal activity. Because interrelations are central to the concept of ecology, we will follow the lead of economics and gauge the size of the system by its total activity, rather than by its contents, or biomass. (This is not to imply that stocks or biomass levels are without importance; they will enter our calculations in the next chapter.) If measuring size in terms of activity seems strange to some readers, perhaps their misgivings will be eased once they are reminded that the familiar "gross domestic product" (GDP) is nothing more than a particular component of the total system throughput. The concept is not so strange after all: just as we recognize a growing economy as one with an increasing GDP, so here we will identify the growth of an ecosystem with an increase in its total system throughput.

4.3   Information Theory in Ecology

The effects of autocatalysis upon the intensive attributes (i.e., development) of a flow network do not lend themselves to quantification in such ready fashion. Let us recall how the pattern of flows is changing: Those pathways that participate most in autocatalysis increase in magnitude, usually at the expense of other processes that are less engaged.

Because autocatalytic loops naturally compete for resources, those links that are less efficient in augmenting autocatalysis are effectively pruned. (It is not that the links necessarily disappear; they simply shrink relative to the favored exchanges.) To put it in other words, if a quantum of material is currently embodied in a particular node, and there are several predators upon that population, transfer is more likely to occur to the predator that will return the most resource to its prey via an autocatalytic route. All else being equal, feeding by the competing predators will shrink relative to the most favored exchange. In effect, the probability for transfer to the favored prey will increase, while those for transfer to the other predators will fall.

The field of mathematics that quantifies such changes in probabilities is called information theory. Myron Tribus has defined information as anything that *causes* a change in probability assignment (Tribus and McIrvine 1971). Before we begin to apply information theory to the process of development in ecosystems, however, I wish to make a brief aside into the nature of information and its prior applications to ecology.

We recall that there has been a not-too-subtle shift in attitude regarding what exactly it is that probabilities measure. This change has been away from the idea that probabilities measure our ignorance about a deterministic situation, toward the notion that they reflect an indeterminacy inherent in the process itself. That is, formerly probability theory could be called quantitative epistemology, but now many acknowledge that it bears as well upon the ontological character of events. Unfortunately, this change in outlook does not seem to have permeated the field of information theory. The central concept in information theory is still called "uncertainty"—a state of knowledge, not a state of nature.

This emphasis upon the subjective in information theory is largely a consequence of its historical origins, which trace back to code-breaking research conducted during the Second World War (Shannon 1948). Most textbooks on the subject are still slanted toward the field of communications. It is understandable, therefore, why so many feel obliged to retain the terminology and perspective of communications engineers. Thus, some think it always necessary to identify a sender, a receiver, and a channel over which information flows. My opinion is that such conservatism is unnecessarily procrustean (Ulanowicz 1986a). Information theory has a legitimacy that transcends the realm of communications theory. Probabilities are the fundamental elements of information theory, and they were around long before Claude Shannon, or even Ludwig Boltzmann, appeared on the scene. In general terms, information theory quantifies changes in probability assignment, in much the same sense that differential calculus quantifies changes in algebraic quantities. Thus, information theory is appropriate anywhere probabilities are germane. Its legitimacy should not be constrained by the peculiarities of its historical development. Lest anyone get the wrong impression, I wish to state clearly that epistemological considerations never can be, nor ever should be, entirely removed from probability and information theories. Nevertheless, I think Popper was on the right track. We always begin work on a problem with some degree of uncertainty. Through repeated observations under different conditions we reduce that uncertainty (gain information). However, under all possible circumstances a residual "uncertainty" will persist due to the inherent indeterminacy in the process and its context.

For this reason, John D. Collier (pers. comm.) has suggested that the term *uncertainty* be replaced by *indeterminacy*. I will try to adopt Collier's terminology simply to emphasize the potential for causal openness in developing systems. To avoid undue confusion, however, I will retain the term *information*, despite its heavy epistemic connotations. It will be used according to Tribus's definition, which hinges more upon the agency behind a change in probability than upon any subjective perception of that change. *Thus, "information" refers to the effects of that which imparts order and pattern to a system.*

I take special pains to extol the universal nature of information theory, because many ecologists, whose bread and butter it is to use probabilities and statistics, nonetheless disdain information theory as an alien art form having no relevance whatsoever to ecology. In part, this prejudice derives from the foreign and confusing terminology in which the theory is usually taught. It is most frustrating for the beginner to encounter the key term *uncertainty* and its ordinary antonym *information* used interchangeably. (I myself confess to having eschewed information theory for years, simply because I rejected the assertion that the "snow" pattern on a TV screen [no signal] contained more "information" than one that conveyed the image of, say, an attractive movie star. Unfortunately, this failure to distinguish the "capacity for information" from information itself persists in too many classrooms, making the subject gratuitously difficult for many students.) Perhaps the most significant reason why so many ecologists reject information theory, however, is rooted in yet another historical accident.

About four decades ago, Robert MacArthur (1955) made a very promising start to introducing information theory into ecology: he used Claude Shannon's formula for indeterminacy to quantify the multiplicity of trophic flow pathways in an ecosystem. MacArthur's purpose was to set up a quantitative test of E. P. Odum's earlier hypothesis (1953) that a redundancy of pathways among the compartments of an ecosystem affords more possibilities for rerouting transfers whenever crucial links are perturbed (figure 4.3). This notion has enormous intuitive appeal. MacArthur served up to ecologists an apparently testable, quantitative whole-community hypothesis: Diversity of flows facilitates ecosystem homeostasis.

Over the next decade and a half, the enthusiasm for testing this hypothesis was phenomenal (Woodwell and Smith 1969). Unfortunately, avenues of testing soon diverged from the outline sketched by MacArthur. Flows in ecosystems are relatively difficult to quantify, so attention soon shifted toward the diversity of more accessible population numbers. Homeostasis proved elusive to quantify, and was permuted into stability as defined in linear analysis. The whole issue gradually devolved into the proposition that "diversity begets stability," cast in rather narrow mathematical terms. Robert May (1973), an erstwhile physicist turned ecologist, brought the whole brittle structure tumbling down by his rigorous demonstration that the existence of more connections among the members of a system is more likely to degrade than to improve its stability. It was a failure of the first magnitude for ecologists, who in their embarrassment turned with a vengeance against information theory. So slow have ecologists been to forgive the culpable theory that several promising reformulations of information for ecology have passed virtually unnoticed (e.g., Atlan 1974; Rutledge, Basorre, and Mulholland 1976).

4.4   Quantifying Information

We now turn our attention back to translating into mathematics the tendency for ecosystems progressively to mitigate indeterminacy. We start with the lead of Shannon and derive a formula for the capacity of a system for either information or indeterminacy. (This capacity for either order or disorder we will take as our surrogate for the system's *complexity*, an otherwise ill-defined and controversial property of systems.) To illustrate how the formula might come about, I begin by describing a game of chance that fascinated me as a small child.

On the corner of the block in my grandparents' neighborhood stood a small confectionery store where I enjoyed spending coins that relatives had just given me. Near the entrance stood a small metal box mounted on a stand with a glass front. At the top of the box was a slot into which

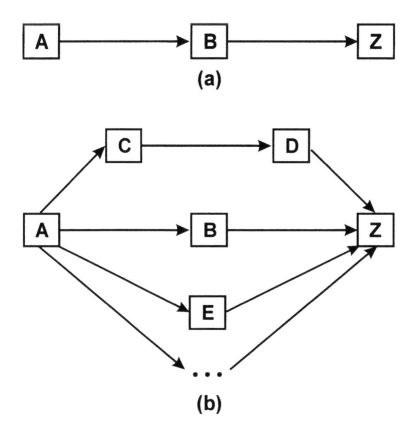

FIGURE 4.3.
(*a*) A highly vulnerable single, indirect pathway of resources from *A* to *Z*.
(*b*) A more reliable configuration of redundant pathways between the same
two elements.

one inserted a penny. The penny dropped through several horizontal
rows of staggered small nails spaced a little more than the diameter of a
penny in all directions; it would fall, bouncing off the nails it encountered and tracing an erratic path, until it settled in one of ten or so slots
at the bottom. If the penny came to rest in a designated slot near the middle, the player received a nickel's worth of candy from the shopkeeper
(figure 4.4).

In our ideal reconstruction of the game, we imagine that the penny
falls through exactly ten rows of nails. Each encounter with a nail slows

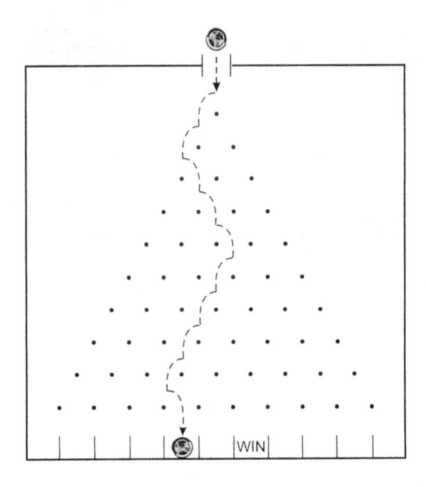

FIGURE 4.4.
A simple game of chance in which a coin, inserted in a slot at the top, falls
through a matrix of nails and lands in one of eleven slots at the bottom (only
one of which designates a win).

the fall of the penny, so that it never has enough momentum to do much
more than fall right or left and encounter the next-lower nail almost
head-on. Thus, we picture a pyramid of nails, one in the first row, two in
the second, three in the third, etc. After encountering the tenth row of
nails, the penny falls into one of eleven slots at the bottom.

Now, if we want to know how complex the game is, we can think of any number of ways of posing the question. We might ask how difficult it is to play the game, for example. (Pretty simple.) But we are seeking to quantify the game's complexity, so we might gauge this by the number of final outcomes—i.e., 11. A good enough answer, but this is a measure of outcome, and we are primarily interested in *processes*, which in ecosystem theory translates into pathways of energy and material flows. Hence, we seek here to enumerate the number of possible unique pathways the penny may take through the system. Because with each encounter with a nail the penny falls either right or left, and because the encounters form a nonrecursive sequence (pennies do not bounce to higher rows), there will be exactly 2^{10} or 1,024 trajectories that are different from each other by at least one nail.

The significant thing to notice is that *the complexity of the system is generated by the number of combinations of possible encounters.* It is well known that such *combinatorics* increase in geometric proportion. In this simple example, we can clearly identify and enumerate the events that generate the complexity (10 encounters with nails). More often, we must be satisfied to count only the combinations themselves, as we do in constructing an events table. To estimate the number of factors that generate these combinations, we must work backward—that is, instead of exponentiation, we must employ the inverse function, the logarithm. Thus we have $\log_2 1,024 = 10$.

In this simple example the logarithmic operation yields the exact number of events that generated the complexity. It was the genius of Boltzmann (and, independently, of Shannon) to perceive the generality of the logarithmic operation whenever it is applied to frequencies or probabilities. Boltzmann postulated that the potential for each configuration to contribute to the overall complexity is proportional to the negative logarithm of the probability that the configuration will occur. The formula expressing this last statement is

$$s = -k \log p,$$

where p is the probability that the configuration will occur, s is its potential contribution to the overall complexity, and k is a constant of proportionality. This formula is inscribed as an epitaph on Boltzmann's tombstone.

If a particular system configuration occurs almost all the time, its probability of occurring is nearly unity. Because $\log(1) = 0$, we see that the almost ubiquitous result does not contribute much to system complexity; it is indicative of a simple system. On the other hand, if a combination occurs only rarely, it has a large potential to complexify matters. A truly complex system will come close to behaving uniquely each time it functions.

In our example of the falling penny there is no reason to assume that any trajectory is more probable than any other. Hence, the equiprobability of any particular sequence, say the *ith* one, is $p_i = 1/1,024$. Recalling that the logarithm of the reciprocal of a number is the negative of the logarithm of that number, and using 2 as the base of the logarithms, we get $s_i = 10k$ from Boltzmann's formula. We just noted that if a particular configuration dominates a system its s_i becomes very small, whereas if a configuration is rare it has a very large potential to complicate matters. However, how much a rare configuration *actually* contributes to the complexity of a system must be weighted by the low frequency at which it occurs. That is, to describe the complexity of a system we need to *average* the potential contributions by states of the system. This we accomplish by weighting each s_i by its corresponding p_i and summing over all states. We refer to this operation as applying Shannon's formula.

Thus far we have discussed only how events can generate complexity in a system. We have not asked how to decide whether these events *inform* the system, i.e., contribute toward a particular ordered pattern, or whether they act to disrupt it by contributing to random, unpredictable behavior. To make this distinction, we refer again to Tribus's definition that information is that which causes a change in probability assignment. In our model of the penny game, we assumed that each encounter with a nail resulted in a 50% chance of the penny's falling either right or left. For the time being, if we confine our attention to a single encounter with a nail, then the two configurations have probabilities $p_i = \frac{1}{2}$, and the average complexity of this subsystem works out to be $1k$.

Now let us suppose that for some reason the probability that the penny will fall to the left (p_1) becomes greater than 50%, and the probability of falling to the right (p_2) becomes correspondingly less. Without loss of generality, let us say that the probability of falling left becomes 70% and that of going to the right drops to 30%. Exactly how this asymmetry comes about we need not specify. It could be, for example, that the cross-section of the nail shafts is oval rather than perfectly circular, and the orientation of the irregularity with respect to the direction of fall is such that most encounters roll off to the left. It could be simply that the entire array is not exactly plumb, that the left corner is slightly lower than the right. Whatever the actual mechanism, this change in probability assignment reduces the complexity of the system, as calculated by Shannon's average, by $0.119k$ bits. (A *bit* is the measure of information inherent in a single binary decision.) Because the first situation was more indeterminate than the second, we conclude that the *information* engendered by whatever caused the bias is $0.119k$ bits.

4.5 Information and Flow Networks

MacArthur had perceived, and now we too begin to discern, how Shannon's formula might be useful in quantifying the status of a trophic flow network. But our perspective is a bit different from MacArthur's: we are seeking to quantify those factors that help constrain flows along certain preferred pathways. Take, for example, the flow through the striped bass compartment (#33) in figure 4.2. An estimated 30.4 mg/m^2 of carbon was consumed by the striped bass population during 1985; this amount either was respired (9.9 mg/m^2/y), became detritus (16.0), or was removed by the striped bass fishery (figure 4.5a). Various unspecified constraints must be at work to give rise to this particular distribution. Were they not operative, random or equivocal allocation of the consumption would have resulted in 10.13 mg/m^2/y of carbon being lost by striped bass to each sink (figure 4.5b). This equivocal configuration yields an indeterminacy of $1.585k$ bits according to the Shannon formula. Whatever the constraints that partition the flows in the observed proportions, they reduce this indeterminacy, or inform the system, by $0.163k$ bits.

From this example, it might seem that one way to measure the constraints inherent in the entire network of ecosystem flows would be to perform Shannon's calculations as we just did on the inputs and outputs around each compartment, weight the result for each compartment by the throughflow of that node, and average over the ensemble. Calculating in this manner probably would underestimate the magnitude of constraints that order the flows. For example, we never asked why there are three flows, and not two or four. For that matter, why is every compartment not communicating directly with every other compartment? If that were the case, we would have had to deal with thirty-seven putative flows out of the striped bass compartment. The difference between the indeterminacy over thirty-seven pathways and the known distribution over three routes would have yielded a higher value for the difference that characterizes

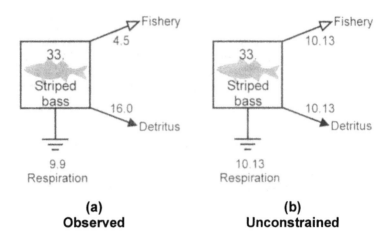

FIGURE 4.5.
(*a*) Example of a naturally constrained distribution of flows out of the striped bass compartment (see figure 4.2). (*b*) How those flows would most likely be distributed if the natural constraints were removed.

the constraints informing the system. Therefore, we seek a calculation that will be inclusive of all constraints at work in the system.[1]

Having introduced the fundamental concepts, however, there is no need to complicate the story any further by belaboring the details of calculation (those wishing to scrutinize the details are referred to Hirata and Ulanowicz 1984, and Ulanowicz and Norden 1990). Suffice it to mention that we define as the most indeterminate network of flows one in which each compartment contributes to and receives flow from all compartments in proportion to its fraction of the total system throughput. Thus, for example, the phytoplankton (compartment #1) in figure 4.2 comprise 0.127 of the total system throughput, and the zooplankton (#8) comprise 0.022; then, under maximally indeterminate conditions, 0.00279 of the total system throughput would flow from the phyto-

---

1. I should point out here the information hidden in the example of the penny falling through the lattice of nails: gravity, which is always pulling the penny in the same direction, and the particular spacing of the nails also are constraints upon how the penny can act. A penny inserted with the same starting momentum into the slot of the same game played in outer space would result in an entirely different set of probabilities, pathways, and complexities.

plankton to the zooplankton. What we observe, however, is 37,149 $mg/m^2/y$ of carbon, or 0.0090 of the TST, taking the route from 1 to 8. We then weight this *change in probability assignment* by the fraction of TST comprised by the actual flow (0.009), and do the same thing for each flow in the network. Summing all these contributions results in what is called the *average mutual information* (AMI) of the flow structure. *AMI measures the average amount of constraint exerted upon an arbitrary quantum of currency in passing from any one compartment to the next.* For the network in figure 4.2, the AMI works out to 2.088 bits.

As shown in the example on striped bass, informed patterns of flow follow a narrower range of pathways than the equiprobable, presumably those that are more efficient in some sense, such as competition or autocatalytic activity. Such "flow articulation" can be represented schematically as in figure 4.6. All three flow networks have the same total system throughput (96 units). Configuration 4.6a is maximally indeterminate: medium at any node is free to flow to any other node (itself included); its average mutual information equals zero. The network in figure 4.2b is more informed: medium currently in one compartment can flow to only two of the other compartments; its AMI is $1k$ bits. The pattern in figure 4.6c is entirely constrained (determined): medium in any compartment must flow to the next in the cycle; its AMI is maximal at $2k$ bits.

Similarly, we might regard the events in tables 3.1 and 3.2 as transfers between compartments. (We are not dealing here with a closed system, because the number of rows does not equal the number of columns.) We would then calculate that the development from the first configuration to the second involves an additional $1.04k$ bits of information per unit of total system activity.

4.6   Ascendency

It appears that we now are able to quantify both the growth and the development induced in a network by indirect mutualism. But we have seen how the two effects can derive from a single or unitary agency. It would be helpful if we could combine our measures of the two effects into a single index. Fortunately, the analytical form of information indices allows us to combine our measures in a very natural way. We have been careful to retain the scalar constant, k, in all measures of information. (Boltzmann himself used this constant to scale his probability measures by the physical dimensions of molecular interactions. His value of 1.3805×10^{-16} $erg/°K$ relates the kinetic energies of molecules to macroscopic measures

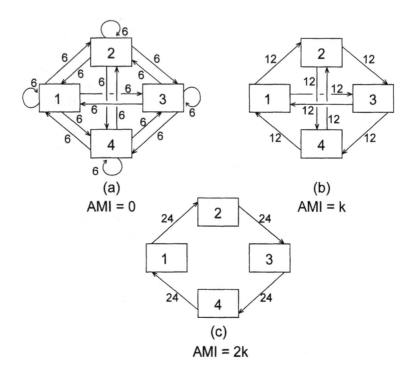

FIGURE 4.6.

(*a*) The most equivocal distribution of ninety-six units of transfer among four system components. (*b*) A more constrained distribution of the same total flow. (*c*) The maximally constrained pattern of ninety-six units of transfer involving all four components.

of temperature. It is considered a fundamental constant of nature that can be found in virtually any table of physical constants.)

When information theory emerged early in this century, any ties of information to physical reality usually were ignored. One simply chose a base for the logarithms, set $k = 1$, and called the ensuing units "bits," "napiers," or "hartleys" according to whether the logarithmic base was 2, *e*, or 10, respectively. We heed Tribus's advice on how to scale information measures by identifying k with the total system throughput. *The resulting product of TST times the network AMI we call the "ascendency" of the system.*

For future reference, we note here that the algebraic result of scaling the AMI by TST is that ascendency now appears as the sum of exactly *nf* terms, where *nf* is the total number of nonzero flows that comprise the system network. Each such term consists of the product of an individual T_{ij} times the logarithm of a complicated expression that involves that particular flow, total inputs to *j*, aggregate outputs from *i*, and the total system throughput. For convenience I will refer to this logarithm as the "logarithmic factor" corresponding to T_{ij} (see Ulanowicz and Norden 1990).

There are a number of agents other than indirect mutualism that can cause the total system throughput or the network mutual information to rise or fall. Over the long run, however, these causes impart no preferred direction to the system. They are as likely to decrease TST and information as to increase them. Indirect mutualism is special in that it possesses an inherent asymmetry of probabilities (a propensity), which works to abet both system size and organization. Furthermore, we have seen how this mutualism arises *at the focal level* and is not simply a passive result of events occurring elsewhere in the hierarchy. These observations lead us to formulate a core hypothesis for developing systems:

In the absence of overwhelming external disturbances, living systems exhibit a natural propensity to increase in ascendency.

There are three significant words pertaining to the statement just made. Two are noteworthy by their presence, the third by its absence. The word *overwhelming* qualifying *disturbances* implies that living systems are always subjected to disturbances. In fact, I argue below that some minimal disturbance is required in order for a system to continue developing. Second, that the statement is statistical in nature is revealed by the use of the word *propensity*, *sensu* Popper. Disturbances may temporarily decrease system ascendency, but the underlying and unremitting restorative tendency is in the direction of increasing ascendency. The only disturbances excluded from the hypothesis are those that overwhelm the autocatalytic configuration itself, i.e., "kill" the system.

Missing from the central postulate is the word *maximize*. Of course, most measures that rise, but remain finite, do eventually approach a maximum. That is not to say, however, that such a maximum exists at the outset as a goal or an objective toward which the system is driven, as can in fact be said for many physical systems. If I were to throw the pen with which I am writing into the air, and if I knew its mass and initial translational and rotational momenta, I would be able to formulate a function, called the Hamiltonian, that could be quantified for any putative pathway. The pen cannot follow just any of these hypothetical trajectories: it is constrained to follow only the one possessing the extremal value for the Hamiltonian function. If I could compute the Hamiltonian quickly enough, the extremum (goal) could be ascertained before the pen completed its trajectory.

Many mechanical systems follow extremal principles in this sense. They are rigidly constrained to maximize (or minimize) certain functions at every step of the way. Ernst Mayr (1992) calls such system behavior "teleomatic." By comparison, however, the telos of living systems is far more plastic and diffuse. Endpoints are not necessarily preordained, and the approach to maximum is not tightly constrained at each instant. At times the increase might even seem desultory. The ecologist Peter Allen (pers. comm.) has described the variational behavior of living systems as akin to "localized hill climbing"; that is, the system responds only to local topography and has little or no information as to the position of any nearby summit. To his description we should add that the system does not always move in the direction of steepest ascent, nor, for that matter, is it tightly constrained to a surface.

If the direction of system growth and development is so loose, then why use the word *ascendency* to describe the system's status? The primary meaning of ascendency is that of domination or supremacy. But it is difficult, though not impossible, to think of ecosystems as competing with one another—so whence the domination? The answer lies in the example of the tossed pen and its associated Hamiltonian. The pen's actual trajectory is superior in a definable and quantifiable sense to any other *virtual* nearby pathway. In like manner, the actual developmental pathway of an ecosystem can be compared with any nearby virtual or alternative configuration. A second reason for choosing *ascendency* is that its root, *ascend*, connotes the upward, or progressive, direction of living systems as they develop from the inchoate into the mature.

Before pressing on, it should be reiterated that the origins of the ascendency hypothesis were strictly empirical in nature. It is presented here as the culmination of a series of definitions and heuristic arguments centered on autocatalysis, as though the notion were deduced from first principles. The concept of increasing ascendency originated, however, in purely phenomenological fashion. In a very seminal paper, Eugene Odum (1969:265) listed twenty-four attributes of ecosystems that might be employed to differentiate whether they were in the early or late stages of succession (table 4.1). Odum's properties can be aggregated according to how they characterize major tendencies, such as those toward greater species richness, stronger retention and cycling of resources, and finer trophic specialization. It was the chance discovery that each of these trends was a separate manifestation of increasing mutual information in trophic networks that first motivated the formulation of ascendency (Ulanowicz 1980). From its very beginning, ascendency was conceived for the purpose of unifying disparate, empirical observations on how ecosystems grow and develop.

4.7   Limits to Ascendency

Like all finite processes, the progression of growth and development cannot continue indefinitely. For any real system the rise in ascendency must eventually slow and possibly even retrogress. What puts the brakes on increasing ascendency? Intuitively, we know at least part of the answer: as systems become more like machines, they perforce lose some of their adaptability and grow increasingly vulnerable to chance new disturbances. But information theory allows us to address the question in a more quantitative fashion, as I shall now outline (cf. Conrad 1983).

Recall that we began our discussion of information theory not by addressing information outright, but by considering instead the *capacity* of a system for either information or indeterminacy. We used Shannon's formula to quantify the complexity of the system in its most indeterminate configuration. Thereafter, we identified information as anything that constrains the system elements so as to change their probability assignments away from the values they have in the most indeterminate state. The amount by which the complexity in the indeterminate state is diminished by the constraints became the information value assigned to those constraints.

Another way of saying the same thing is to recognize that the initial complexity consists of two components—an ordered fraction and a disorganized residual. From this perspective we scaled the information measure (the component representing order or constraint) by the total system throughput and called the result the system ascendency. Let us now scale the initial complexity and the residual term (representing disorder or freedom) by the same total system throughput, and call the results the *development capacity* and the system *overhead*, respectively. In this new vocabulary, the development capacity can be expressed as the sum of the organized ascendency and the still-indeterminate overhead. Equivalently, the system ascendency becomes the difference between system capacity and overhead (figure 4.7). This latter relationship allows us to divide the question of what limits the increase in ascendency into two complementary aspects: (1) What limits increases in system capacity? and (2) What keeps overhead from disappearing?

To address the first question, I repeat that system capacity is the product of the total system throughput times the Shannon diversity of the
individual flows (Ulanowicz and Norden 1990). The TST of any system
will remain finite so long as the exogenous inputs to the system remain
bounded. Although it is true that TST can increase (via recycling) even
when aggregate inputs are fixed, the second law requires that some currency be lost (dissipated) on each pass through a compartment; hence,
the increments to the TST via recycling diminish with each successive
time around a loop. This attenuation guarantees that the sum of all passes
through all cycles must converge to a finite quantity. Finite inputs *guarantee* a finite TST.

TABLE 4.1

A Tabular Model of Ecological Succession: Trends to be Expected in the
Development of an Ecosystem (after E. P. Odum 1969:265)

	Development stages	*Mature stages*
Ecosystem attributes		
1. Gross production/community respiration (P/R ratio)	Greater or less than 1	Approaches 1
2. Gross production/standing crop biomass (P/B ratio)	High	Low
3. Biomass supported/unit energy flow (B/E ratio)	Low	High
4. Net community production (yield)	High	Low
5. Food chains	Linear, predominantly grazing	Weblike, predominantly detritus
Community structure		
6. Total organic matter	Small	Large
7. Inorganic nutrients	Extrabiotic	Intrabiotic
8. Species diversity—variety component	Low	High
9. Species diversity— equitability component	Low	High
10. Biochemical diversity	Low	High
11. Stratification and spatial heterogeneity (pattern diversity)	Poorly organized	Well organized

Even though TST may have reached its limit, this does not mean that the second factor comprising the capacity cannot continue to increase. The Shannon diversity of flows is calculated using the probabilities of each individual flow in the context of total system activity. That is, from among all flows occurring in the system at a given instant, what is the probability of selecting a quantum of currency that is passing from compartment i to component j? This probability is readily estimated by the quotient $T_{ij}/$ TST. All such probabilities then go into Shannon's formula to yield the diversity of flows (the second factor in the development capacity).

Whenever the same amount of total throughput is partitioned among a greater number of flows, the Shannon index will increase. To main-

TABLE 4.1 (CONTINUED)		
Life history		
12. Niche specialization	Broad	Narrow
13. Size of organism	Small	Large
14. Life cycles	Short, simple	Long, complex
Nutrient cycling		
15. Mineral cycles	Open	Closed
16. Nutrient exchange rate, between organisms and environment	Rapid	Slow
17. Role of detritus in nutrient regeneration	Unimportant	Important
Selection pressure		
18. Growth form	For rapid growth ("r selection")	For feedback contributions ("K selection")
19. Production	Quantity	Quality
Overall homeostasis		
20. Internal symbiosis	Undeveloped	Developed
21. Nutrient conservation	Poor	Good
22. Stability (resistance to external perturbations)	Poor	Good
23. Entropy	High	Low
24. Information	Low	High

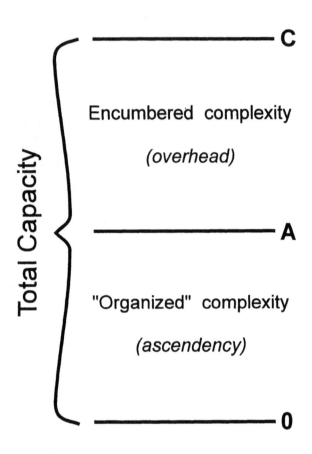

FIGURE 4.7.
Schematic representation of the segregation of development capacity (complexity) into disordered and organized components (overhead and ascendency, respectively). (A = level of ascendency; C = level of capacity.)

tain ever finer partitioning of flows eventually requires that we partition the components themselves. The problem with sustaining arbitrarily many small components is that some of the smallest units invariably will succumb to chance perturbations; hence, the flow diversity cannot rise indefinitely, but must eventually be retarded by the intensity and frequency of perturbations to the system.

To recapitulate: Increases in system capacity are limited by a combination of finite resource availability and the extinction of small system components due to environmental perturbations. The reader might have recognized in the second factor an aspect of biodiversity and the problems associated with maintaining it. In modern times, increases in the amplitude and novelty of anthropogenic stresses have acted as just described to decrease the diversity of components (species) in many of the world's ecosystems. *Information theory then quantitatively relates this drop in diversity to a diminished capacity of the system to grow and develop* (Norton and Ulanowicz 1992).

It sometimes happens that natural systems can plateau in their development capacity but still continue to develop. In such cases the ascendency is increasing at the expense of overhead: such inchoate elements as comprise the overhead become incorporated into organized structure. An overt example is the development of the human autonomous nervous system. Physiologists tell us that the multiplication of nerve cells in the human organism ceases sometime before about 3½ years of age. Because the number of neural cells and connections is maximal at this age, and because electrical activity does not appreciably increase with age, the development capacity of the nervous system does not increase for the remainder of a person's lifetime. Yet we know that the nervous system continues to develop throughout most of a person's remaining years. This development occurs via the increasing definition (articulation) of neural pathways already in existence.

As the nervous system develops during the course of fetal and natal ontogeny, some pathways become "hardwired." These pathways undergird reflex responses. The brain of the young human (especially the cerebellum) nonetheless remains very diffusely connected. In response to stimuli experienced by the individual, certain connections are facilitated, and others inhibited. The neural system develops. Although the overall capacity of the brain remains more or less constant, the fraction of connections entrained by structured responses (ascendency) continues to climb until senility (Ulanowicz 1980).

We now turn our attention away from limits on capacity, toward factors that constrain the decrease of system overhead. Each flow in a network contributes a term to the ascendency and overhead. Earlier, we identified four categories of material and energy flows: (1) inputs, (2) dissipations, (3) exports, and (4) internal transfers. Therefore one useful way of grouping the terms comprising the overhead is according to the types of flow that generate them (figure 4.8). We now consider each of the four groupings in turn and ask what keeps each cluster from disappearing.

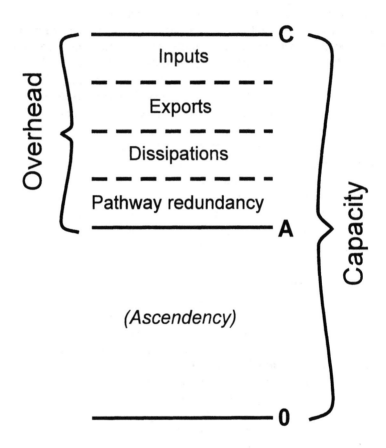

FIGURE 4.8.
Refinement of figure 4.7, showing that system overhead can be decomposed
into distinct fractions, each of which is generated by redundancy in one of the
four fundamental types of flows.

Inputs

The group of overhead terms generated by exogenous imports is
modulated both by the magnitudes of the inputs and by their multiplicity. The magnitude of each import scales its contribution to the overhead.
As with the ascendency, this magnitude multiplies a second (logarithmic) factor. The magnitudes of the logarithmic factors depend upon the
distribution of the input magnitudes, just as in the Shannon index. The

Shannon index is maximal when the distribution is even. Therefore, a given total amount of input will contribute most to the overhead when it is partitioned evenly among those components that receive inputs. Conversely, if that amount is unevenly distributed, its contribution to the overhead will fall. In the particular case where all input flows into a single component, the grouping of input terms in the overhead becomes equal to zero, regardless of the magnitude of the single import.

With this as background, we now may ask: What circumstances will lead to a decrease in the overhead on inputs, and what others would halt this decline? All other things being equal, the overhead terms will fall whenever the magnitudes of the inputs decrease. But we saw a few paragraphs ago how these same magnitudes modulate the TST, which scales the development capacity (and all indices, for that matter). It would be highly counterproductive to system activity, therefore, to curtail the absolute magnitude of its imports. In a manner of speaking, all ships would fall with this ebbing tide.

We saw, however, that capacity can be augmented through a greater recycling of material and energy. That is, when input intensities remain constant but recycling increases, inputs become a decreasing fraction of total system throughput, and development capacity will grow faster than will the overhead on imports. It is not surprising, therefore, that systems often develop in a direction in which inputs become a progressively smaller portion of total activity. Such "internalization" of activity is a favored scenario for economic and corporate growth.

For a given amount of input flow, ascendency will grow at the expense of overhead if those imports enter the system via fewer compartments. Because ascendency is a surrogate for efficiency, specializing in a few inputs means that the system mines those sources from which medium is most readily available and least costly to import. Ceteris paribus, we expect systems to develop in this direction. How *far* they can go in this direction is dictated, however, by the rigors of the environment. The danger of relying too much on one or a few inputs is best captured by the proverb, "Don't put all your eggs in one basket!" A system that is sustained via only a few pathways is highly vulnerable to any disruptions of those lines. We conclude, then, that *systems develop in the direction of more efficient imports along fewer links up to the point where environmental disruptions of those links create the need for compensatory additions from other, less-efficient sources.* As a result of this balance, we would expect a narrower dependence on sources to develop in more benign environments, and a more uniform reliance to prevail in more rigorous surroundings.

Dissipations

Some readers may question why we distinguish between usable exports and dissipative ones. The distinction now becomes a key feature in elaborating the limits to system development. First, we consider the contributions made to the overhead by the dissipation of energy and (in a sense) material. As with the overhead on inputs, the total encumbrance due to dissipations depends both upon the magnitudes of those losses and on their distribution. Both factors are driven by the second law of thermodynamics. All real processes result in some dissipative loss. Any change that would increase the efficiency (decrease the dissipative loss) of a component would, of course, be favored under a scenario of increasing ascendency. But such losses can never be reduced to zero.

The extent to which a dissipative flow constitutes a loss depends on where in the hierarchy the observation is made. At the level of the ecosystem or the whole organism, most of the dissipative losses take the form of respirations. We know that most of this dissipation is due to processes that create and maintain order on a cellular or subcellular scale. Thus, when viewed from a microscopic perspective, metabolic losses acquire more the flavor of an investment. This insight allows a more complete statement of why dissipative losses cannot be entirely eliminated: they are not only required by the second law, they also represent an encumbrance necessary to maintain organization at lower scales.

As for the multiplicity and distribution of dissipations, not much can be said at this point without getting too far ahead of our story. The second law mandates that dissipations be widespread in most systems. Deviations from a uniform distribution of dissipations, however, become most apparent at either end of the trophic spectrum. Both bacteria and top carnivores tend to generate higher respirations per unit biomass. Furthermore, dissipation tends not to be uniform in time, but to occur more in pulses (H. T. Odum 1982). Sometimes pulsing is infrequent but very strong, as when ecosystems collapse, or reset themselves for another developmental cycle (Holling 1992). In the next chapter we will learn that pulsing can be a form of temporal structure; as such, it makes proportionately less contribution to the dissipative overhead term.

Exports

If some dissipations represent necessary encumbrances at lower hierarchical levels, we might expect that exports of usable currency generate overhead that is demanded by higher levels. And indeed, this is the case—so that I initially called such overhead *tribute*. Whenever there exists no feedback pathway at the metasystem level whereby exports from a given system can positively affect imports to the same system, then it always improves system development to eliminate tribute to the fullest extent possible. If, however, exports from a system benefit its environment in such a way that the latter can release more resources to the given system, then decreasing such exports could be counterproductive.

It is exactly such feedback in economic systems that currency and credits were invented to promote. It is not surprising, then, that the clearest examples of why exports cannot be arbitrarily curtailed come from economics. In the aftermath of the 1970s oil crisis, for example, oil-producing states cut back on exports of this resource to drive up its price substantially. As a result, the global economy was affected to such a negative extent that it jeopardized the ability of the oil states to obtain the goods they needed in order to develop.

One can easily imagine the same scenario transpiring in an ecosystem. For example, a population of birds might visit an island to consume the seeds of a particular tree species. If the tree receives no return for this export, it is likely to evolve an adaptation to make its seeds less available to the birds; the island ecosystem appears to shepherd its own development by decreasing its tribute. Suppose, however, that during their visits to the island (and especially to the trees) the birds deposited locally rare trace elements or nitrogen in their droppings; under these circumstances, less-attractive seeds would mean fewer birds, without whose droppings the production of the whole ecosystem could suffer. If the coupling (reward) to the trees becomes tight enough, it will even benefit tree species to make their propagules more attractive to the bird, perhaps by enveloping the seed in a nutritive coating (fruit).

Internal Transfers

The final category of flows involves intrasystem exchanges. The overhead terms generated by these flows represent the degree of pathway *redundancy* within the system. For example, the streamlined network in figure 4.6c contains no redundant pathways: the overhead terms generated by the four internal exchanges are all equal to zero. In contrast, all of the development capacity of figure 4.6a appears as overhead (i.e., ascendency is zero). As this hypothetical example has no connections to the external world, all of this overhead represents the redundancy of pathways in the system.

Like other forms of overhead, redundancy imposes a cost on the system. For instance, all indirect pathways between any two points in the network will not be equally efficient at conveying medium. To maintain the same amount of exchange between the points, it will cost more in terms of dissipative losses to send the flow over diverse routes than it would to channel it all over the most efficient route. Furthermore, when using multiple pathways there is always a greater chance that the end compartment will receive the resource out of phase to its own needs, due to the differences in the lengths (and times) of transmission over the different pathways. Hence, the natural tendency during development is for the system to shed its redundant connections. As soon as feedback enters the picture, as it does in neural development, the tendency toward less redundancy becomes accelerated.[2]

Just as a system with only a single input is highly vulnerable, so too one with minimal internal connections becomes highly susceptible to external perturbations. A sustainable system requires a balance between ascendency and redundancy. Exactly where that balance is struck will depend on the rigor of the environment. Stable and benign environments, such as those in tropical rain forests, should permit a larger number of trophic specialists to survive. On the other hand, ecosystems with erratically fluctuating environments, such as estuaries, should contain a larger proportion of opportunistic and omnivorous feeders.

4.8   A Scenario for Development

These considerations on the limits to ascendency allow us now to employ the ascendency, capacity, and overhead to compose a generic scenario for the (I) growth, (II) development, (III) maturation, and (IV) senescence of living systems (figure 4.9). In its beginning, immediately after a system has survived a major destructive perturbation or when it is invading a new spatial domain, its initial response is to augment its activities and biomass at the fastest rate possible (Golley 1974). This means that both capacity and ascendency will increase abruptly, fueled in large measure by a burgeoning TST. Because resources are usually very abundant during this inchoate stage, almost no competitive selection occurs then.

---

2. Robert May (1973) has pointed out that too many connections can destabilize a system. Using information theoretical variables instead of mechanical dynamical variables, it becomes possible to identify a threshold connectivity of about three connections per node, above which the system disintegrates due to internal instabilities (Ulanowicz forthcoming).

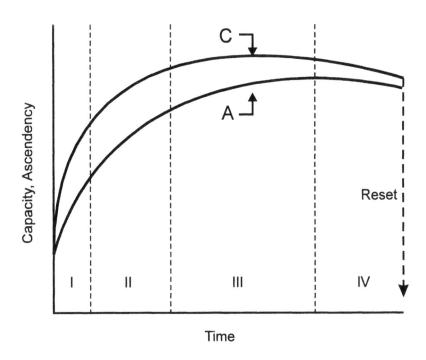

FIGURE 4.9.
Hypothesized trajectories of the developmental capacity (C) and the ascendency (A) over the lifetime of a generic living system. The four phases are described in the text.

In the development phase, even as the TST begins to decelerate due to the initial decline in the relative availability of resources, competitive pressures remain mild, and new immigrant species are seldom excluded. The consequent rise in flow diversity contributes progressively more to the increase in capacity (vis-à-vis that generated by increasing TST), and eventually comes to dominate the final increments in development capacity. The ascendency tracks but lags behind the growing capacity in these first two stages.

Capacity plateaus during the third phase (maturity), and selection pressures increase. As outlined in the example of the nervous system, a pruning of redundant pathways ensues, leaving dominant those routes (and compartments) that most effectively participate in autocatalytic activities. Flow and biomass diversities often decline somewhat during maturity, but ascendency continues its slow ascent at the expense of overhead.

With declining overhead comes a diminished capacity to withstand disturbances. As a result, capacity slowly begins to erode, and with it, ascendency. The system enters its fourth and final stage, senescence. What happens as senescence continues depends upon the particular system and its environment. Organisms usually collapse catastrophically and lose their identities (they die). Some ecosystems similarly exhibit an abrupt disintegration and revert to the immature stage. Such collapse can take the form of a massive canopy fire, as happened a few years ago in Yellowstone National Park, or of violent outbreaks of pests, such as the spruce budworm in the coniferous forests of the Pacific Northwest (Holling 1986).

Virtually every ecosystem ecologist has his or her own rendition of this development scenario. Crawford Holling (1986), for example, emphasizes the cyclical nature of ecosystem development. He portrays ecosystem change as a sequence of four "functions" that follow one another in cyclical fashion. The initial activity is the *exploitation* of released nutrients by fast-growing species. This function blends slowly into a *conservation* phase that is dominated by slower-growing, bio-mass-accumulating populations. The latter become victims of their own success, as these major system elements become linked tightly to each other along obligatory pathways. Such "brittle" configurations become exceedingly vulnerable to external perturbations, and the eventual result is a *catastrophic* collapse or resetting of the system—what Holling calls "creative destruction." There follows in the final phase a release of nutrients from dead material, or the process of *renewal*, which makes possible again the exploitative phase. All transitions are relatively rapid, save for the slow succession from exploitation to conservation.

Holling portrayed his scenario graphically in what is fast becoming an icon of ecosystem science—his "figure eight" diagram (figure 4.10). In this depiction, the capital stored by the system is plotted against the system's organization, which Holling relates vaguely to system connectedness. As the ecosystem proceeds through its cycle, it traces out a self-intersecting curve, much in the likeness of the number 8 laid on its side.

As mentioned, Holling's exemplar is the spruce budworm cycle common to the coniferous forests of the Pacific Northwest. During such an outbreak much of the system biomass is lost via tree defoliation, but a significant fraction remains standing as boles (tree trunks), and this capital is released only slowly in the process of renewal. Hence, one wonders exactly why Holling's figure takes the downward turn at the right-hand side of his figure. Why should there be a sudden loss of capital without a concomitant loss of organization? Why does the system not turn

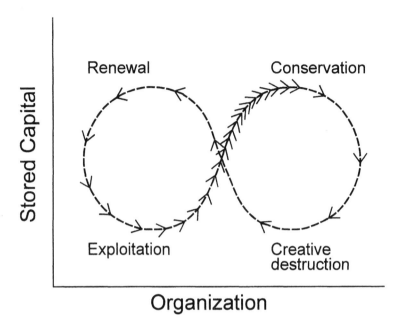

FIGURE 4.10.
Crawford Holling's suggestion (1986) for how ecosystems progress through
the sequential phases of renewal, exploitation, conservation, and creative
destruction. When the stored capital of such a system is plotted against some
measure of its organization, a figure eight is traced on the phase plane. The distance between two successive arrows represents the relative speed at which the
system is traversing the loop during that interval. (Adapted with permission.)

suddenly from the upper right-hand corner back toward the vertical axis,
giving rise to a triangular pattern—or perhaps fall back abruptly along
the same diagonal line that it slowly ascended during succession?

These questions prompt an attempt to plot the ascendency scenario on
a phase plane very similar to the one that Holling used. We can, for
example, roughly equate stored capital with system biomass (or the
accumulated amount of some particular element, as the case may be). In
place of "organization" or "connectedness" we might prefer a quantitatively more explicit measure—e.g., the average mutual information of
the trophic connections.

We also note a difference in opinion as to what constitutes the renewal
and exploitation periods. Holling identifies renewal as the breakdown of
biomass by biological and physical agencies that slowly releases nutri-

ents. Renewal in our narrative is assumed to follow perturbations, like fire, that very suddenly destroy both organization and biomass, releasing nutrients in the process. Thereafter, the system renews itself according to Frank Golley's script (1974), wherein surviving biota quickly capture the newly released abiotic resources.

We begin our plot, as did Holling, in the lower left-hand corner of the diagram. During renewal biomass increases suddenly and relatively independently of system connections. That is, the system trajectory rises sharply, almost parallel to the vertical axis (figure 4.11). This increase in biomass necessarily slows as the abiotic resource pool is diminished, and the new biomass itself becomes available for exploitation by a

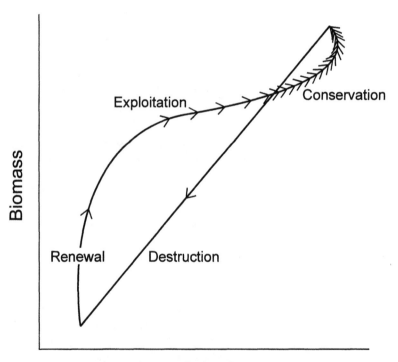

Mutual Information of Flow Structure

FIGURE 4.11.
Hypothetical counterpart to Holling's diagram in figure 4.10, created by plotting the biomass of the system against the mutual information inherent in its flow structure.

greater variety of biotic players. Hence, during exploitation the curve makes a hyperbolic turn to the right.

Movement more closely parallel to the horizontal axis continues during the beginning of the mature phase, as pruning that favors the more efficient and conservative trophic pathways takes place against a background of only slowly accumulating biomass. Eventually, what restructuring is possible will run its course, and further increase in organization must slow. The accumulation of biomass, however, may continue during the later mature (or conservation) phase at its slow but steady pace, causing the system trajectory to bend upward. As the system slowly rises, it may even begin to bend back toward the vertical axis (cf. the decline of *C* and *A* in figure 4.9). Somewhere around this point the "brittle" system may succumb to catastrophe, in which case the trajectory returns very abruptly along a diagonal to the vicinity of its starting point. The resulting plot resembles a forward-leaning and incompletely drawn G clef from musical notation.

In comparing figures 4.10 and 4.11, we can immediately conclude that they are topologically equivalent. (In mathematical topology, two forms are considered to be equivalent if they can be made to conform merely by stretching them in places.) One obvious difference is that the rotational sense in which the patterns are traversed is opposite in the two cases. This discrepancy owes to two factors: First, renewal is interpreted differently in the two descriptions; Holling's renewal pertains to subboreal terrestrial ecosystems, whereas the ascendency perspective derives more from the dynamics of marine, estuarine, and subtropical "fire climax" communities. Second, the quantification of organization via the mutual information inherent in the trophic structure allows one to follow meanders in the trajectory more closely. On a fundamental level, however, the stories and pictures overlap considerably.

In spite of such congruence, we may still question Holling's insistence upon large-scale system surprise. Certainly, it is a key feature of many ecosystems. But other systems—many of which are nearer equatorial latitudes, such as tropical rain forests or ocean gyre ecosystems—appear to persist for very long periods while exhibiting only minor oscillations in the upper right-hand corner of Holling's diagram. For instance, some ecosystems have evolved to mitigate the trend toward senescence by incorporating small "overshoots" into their behavior. An example is the xeric mesophyte, or scrub growth, that once covered most of the porous sandy soils in central Florida: vegetative litter and stems build until the potential for fire is finally realized; recovery is from roots and fire-adapted propagation. Even faunal denizens of the scrub become obligately tied to this cycle of overshoot (Wolfenden and Fitzpatrick 1984). Another fire-adapted system is the High Pine ecosystem (Myers 1990). Is it possible that Holling's focus on "creative destruction" owes more to the biogeographic circumstances surrounding his research, or perhaps to his predilection for Joseph Schumpeter's emphasis on catastrophic changes in economic and social systems (Elliot 1980)?

4.9   Overhead, Adaptability, and Evolution

Whether or not one views large-scale catastrophe as necessary to the scheme of things, the debate does point our attention toward a much-neglected aspect of system development—that of senescence (Salthe 1993). In particular, the endpoint of senescence, owing as it does to insufficient overhead, engenders in us a new appreciation for the *necessary* role that inefficient, incoherent, redundant, and ofttimes stochastic events and processes play in maintaining and even creating order throughout the lifetime of a system (Conrad 1983). Our human inclination is to seek an ever more orderly and efficient world—which is only natural, considering the degree of chaos and mayhem that characterizes human history. But our intuition tells us that there also can be too much of a good thing. We often speak of individuals' lives and whole societies that are too rigidly structured as being "suffocating." As we have seen, ecosystems, too, can create too much structure and thereby become "brittle." Thus, efficiency can become the road to senescence and catastrophe.

So long as the magnitudes of perturbations remain within certain bounds, however, and occur on a more-or-less regular basis, ecosystems will develop. That is, their ascendencies tend to increase through the pruning of their less efficient, less cooperative elements. But when a system is confronted by a novel or extremely infrequent challenge, something that under normal circumstances had been a liability suddenly takes on a potential for strength-in-reserve. It is from the reservoir of sundry and unfit processes that comprise its overhead that the system draws to *create* an adaptive response to the new threat.

This glimpse into the origins of creativity in the natural world highlights an often unappreciated facet of human creativity. We correctly assume that to be creative a person requires a given amount of the right mental "machinery" or intellectual capacity—but it happens all too often that the brightest individuals are not that creative. Many (I am prone to argue *all*) acts of human creation involve mistakes, chance happenings, or misconceptions that occur at or prior to the moment of creativity.[3] Failings, it seems, are necessary adjuncts to creativity in the human domain, just as they are in the natural realm (Norton and Ulanowicz 1992).

All of which brings us back to our central theme of the nature of causality in the natural world. In a Newtonian world of only material and mechanical causes, creation simply has no place. Perhaps creation happened at some distant time, or occurs at some remote scales, but in the here and now everything runs smoothly according to the laws of physics. Creation in the realm of everyday experience is but an illusion. Against this scenario, C. S. Peirce and Karl Popper have argued that mechanical and material causes are insufficient to account for the order we see about us. Rather, we need to reevaluate older conceptions of cause or formulate new ones, in order to extend the domain of causality over the full range of phenomena that we observe (Ulanowicz 1980).

Now the crux of what Popper (1990) has said finally confronts us full force. He argues that we cannot extend this blanket of causality to cover all phenomena, no matter how hard we may try. There will always be some holes in the fabric. Furthermore, he maintains, it is *necessary* that the world remain open if life and evolution are to be at all possible. Unless we accept the world as open, Popper claims, we can never achieve a true understanding of change in living systems, or achieve significant insights into the nature of life itself.

4.10   A Two-Tendency Universe

Perhaps an open world is not the only enigma to which our discussion is pointing. We also seem to inhabit a world of opposites. We have seen how the mathematics of information theory allows us to dissect system behavior according to its ordered and disordered attributes: ascendency represents how efficiently a system operates; overhead is the catchall for its inefficiency (but it also encompasses, among other things, its reliability). Whenever a system's development capacity remains constant, any increase in one attribute implies a decrease in the other. There is a

---

3. I vividly recall how I stumbled upon the beginnings of ascendency. I was reading an article by Henri Atlan (1974) and had mistaken the sign of a particular term in one of his equations. The term represented the concept of redundancy, which appears in information theory in a guise entirely opposite to the way it has been used in this chapter. My mistake in reading set into motion an entire complex of events that has culminated in this book.

fundamental incompatibility between the ordered and disordered fractions—yet they are complementary aspects of what is essential to sustaining the operation and persistence of the system.

The conflict between orderliness and stochasticity is hardly unfamiliar to systems ecologists. Both Joel Hagen (1992) and Frank Golley (1993), for example, have emphasized the fundamental cleavage of the ecosystems research community into holist and probabilist camps. The holists lay stress upon a perceived organism-like order that presumably originates at the macroscale. Their roots stretch back to Frederic Clements (1916) via G. Evelyn Hutchinson and Eugene Odum (1959). One branch of holists prefers to regard ecosystems less as organisms and more as machines that derive their order from fundamental physical principles (e.g., Lindeman 1942; H. T. Odum 1971). Arrayed against the holists are the probabilists, who stress the indeterminate nature of events at the hierarchical levels of the population and below. Their pedigree traces to Henry Gleason (1917), who described the formation of plant communities in terms of a stochastic, and therefore unpredictable, scenario. George Williams (1966) and Daniel Simberloff (1980) subsequently have elaborated the probabilist theme using as background the neo-Darwinian synthesis. One can almost envision the holist camp unfurling the banner of ascendency, while their opponents respond by showing the colors of overhead.

The antagonism between ascendency and overhead extends as well to the separate tendencies that each index exhibits. Our central hypothesis posits a natural movement in the direction of increasing orderliness, which diametrically opposes the omnipresent trend toward decay and dissipation demanded by the second law. Some maintain that this tension is illusory and will vanish once the second law has been reformulated in a more universal way (E. D. Schneider and Kay 1994a). But most are inclined, through familiarity with what has been called the "human condition," to regard this basic tension as more than an illusory quirk of perspective. Perhaps the relationship between ascendency and overhead is best characterized as a "nonagonistic" conflict (David Depew, pers. comm.). Under most circumstances, there is a clear conflict between the two attributes—an increase in one implies a decrease in the other. During those (relatively rare) times that the restraints on rising capacity are absent, or retreat, the conflict between ascendency and overhead abates. At all times, however, any living system requires some proportion of both attributes to survive. In the same loose sense that the centripetality engendered by positive feedback provides a precursor for selfishness and ego, the relationship between overhead and ascendency prefigures a human dialectic (Salthe 1993). Conflict at one scale can turn mutualistic at the next-higher level.

The directions our discussion is taking lead to further intriguing possibilities, but I will postpone treating them until the last chapter. For now, I will focus on extending the concept of ascendency to cover a wider range of natural phenomena.

5

EXTENDING ASCENDENCY

When we think about or investigate ecosystems, we immediately confront elaborate detail. We are attracted not only by the great variety of plants and animals, but also by the very visible way in which habitats are juxtaposed in space. Further, we are intrigued by the frequencies with which biological events wax and wane, usually in connection with regularities in their physical surroundings. Thus far, none of these considerations has entered our discussion.

The image of ecosystems that we have implicitly used thus far derives from chemical reaction theory, where the assumption is that processes occur in a "homogeneous stirred-tank reactor." In this analogy, species are uniformly distributed over the entire habitat, and the rates at which individual organisms encounter one another are the same at all points in the ecosystem. For a system that is large and contains much the same diversity of habitats over its entire extent, these are not bad assumptions. It happens more often, however, that some species aggregate in certain spatial subdomains that do not overlap uniformly with the ranges of other populations in the system. Under these more general conditions the spatial distributions of the populations become important modulators of trophic exchanges. In this chapter we will consider these added dimensions of complexity.

5.1   Spatial Heterogeneity

Actually, spatial heterogeneities do not alter our formal treatment of events. They simply increase the quantity of data necessary. Those familiar with the analysis of physical or biological processes that vary over space have probably encountered the technique of dividing the entire spatial domain into an array of much smaller spatial segments. Often such division is into a "gridwork" in one, two, or three dimensions. Segmentation is usually, but not necessarily, rectilinear; but what is important is that the space be divided into a finite, and therefore countable, number of segments. As long as the segments can be counted, only a single index is needed to identify any particular spatial segment. Thus, although it is usually convenient to number a grid in two or more dimensions using several indices (e.g., identifying a segment in a 2-D spatial array by its row and column location), in principle, a single index will suffice. For example, one begins counting across the first row of the array using as many integers as needed, then drops to the beginning of the second row and continues counting (table 5.1). In an array of n rows of m elements each, every segment would be assigned a unique integer from 1 to mn. This counting method immediately generalizes to three dimensions. It is the same scheme used to store the elements of a multidimensional array in a block of computer memory.

In an ecosystem landscape, the network of trophic exchanges extends not only over species, but over spatial segments as well. That is, we usually speak of a quantum of medium incorporated in species i currently in segment p, moving into population j found in section q. Often the transfer will be simply from i in p to j in p (a feeding event in p), or from i in p to i in q (the migration of a member of species i from p to q). Notationally, we could let T_{ipjq} represent the transfer of species i in segment p to population j in location q.

Closer inspection reveals, however, that only two and not four indices are required. For, just as we can identify uniquely the segments in a two-

TABLE 5.1.
Scheme for Consecutively Numbering the Elements of a Two-Dimensional, $n \times m$ Array

	1	2	3	4	...	m
1	1	2	3	4	...	m
2	$m + 1$	$m + 2$	$m + 3$	$m + 4$	...	$2m$
3	$2m + 1$	$2m + 2$	$2m + 3$	$2m + 4$	...	$3m$
4	$3m + 1$	$3m + 2$	$3m + 3$	$3m + 4$	...	$4m$
...	...	...	...	...	...	...
n	$m(n - 1) + 1$	$m(n - 1) + 2$	$m(n - 1) + 3$	$m(n - 1) + 4$	...	nm

or three-dimensional grid using but a single index, so in the case just discussed, we can devise a convenient numbering scheme that associates a unique integer with each combination of species i in segment p. That is, we can talk just as before about the transition T_{ij}, where i and j run not merely over the number of species in the ecosystem, as in the last chapter, but over the entire gamut of (countable) combinations of species and spatial segments. The upshot is that by a judicious numbering scheme we are able to introduce spatial dimensions into all previous definitions of ascendency, capacity, overhead, and their various components without altering their algebraic forms. The range of the indices and the dimension of the T_{ij} array do, however, increase dramatically, reflecting the fact that a full quantification of the system becomes an arduous exercise in data collection.

5.2   Temporal Changes

In one sense, accounting for temporal changes in ecosystem components is simpler than bookkeeping over spatial domains, but algebraically it becomes a bit more complicated. The simplification derives from the fact that time is unidirectional. Whereas movement in space can be in any direction, progression in time always proceeds from one "instant" to the very next—from time k to time $k + 1$. Any other transitions are nonsensical.

As a result, it makes little sense to compound time with species in the same way that we just folded spatial elements and species into a single index. This is because time is already implicit in the transition from species i to species j. It would be wasteful to define four indices and speak about the transition from species i at time k into species j at time l, because the only temporal transition that makes any sense is when $l = k + 1$. Thus, to make time more explicit, we define T_{ijk} as the transition from species i to component j during time interval k.

Unfortunately, the algebra involved in calculating information over three dimensions is more cumbersome than that necessary for two.[1] I refer the interested reader to Claudia Pahl-Wostl (1992, 1995) for full details on computing ascendency over long time series. What is intriguing about temporal ascendency analysis is that, just as the overhead could be decomposed into meaningful groupings of terms, so certain clusters

1. Here I am referring to the algebraic dimensions connoted by the three indices i, j, and k, not to the three dimensions of physical space.

of terms in the temporal ascendency can be associated with regularities (i.e., frequencies) in the inputs, internal transfers, and outputs. These identifications make it possible to follow the appearance of endogenous frequencies within systems, or the adaptation (entrainment) of system frequencies to identifiable driving signals in the inputs.

We come to see that the applications of ascendency and the components of overhead are not limited to systems that are uniform in space and time. Each spatial order and every temporal coherency that is observed in the natural world can be quantified as a contribution to system ascendency. In like fashion, disorder over space and time can be characterized and quantified by clusters of terms appearing in the generalized system overhead. Again, the only difficulty in rendering a full accounting of the ascendency and overhead of heterogeneous systems is the large *volume* of data required.[2]

5.3   The Influence of Stocks

Thus far, our discussion of systems behavior has been predicated solely upon processes. We have been following the lead of economic theory, which places primary emphasis upon *transactions* over and above any relevance that the contents of sectors may have for system functioning. In my first book I argued, as I did above in chapter 4, that flows alone will suffice to make a quantitative *description* of system behavior (Ulanowicz 1986a). The effects of the compartmental contents were assumed to be written into the network of flows. The purpose of this essay, however, is to venture ever so cautiously beyond mere phenomenological description and to begin to address causality. We know from both experience and intuition that the size and natures of the donor and recipient can strongly influence any exchange between compartments. We seek, therefore, to make at least some characteristics of the system components explicit in the calculation of ascendency.

In economic systems, the capital inventory of participating agents is considered to influence transactions between sectors or corporations. Likewise, in ecosystems analysis the stock of biomass is usually the first property of a population that investigators assess. I have long sought to incorporate the biomasses of compartments into the expression for sys-

2. This is qualitatively quite different from, and more feasible than, the arbitrary *precision* of data that reductionists require in order putatively to model higher-order phenomena.

tem ascendency. Only recently have I been able to introduce it in a way that is dimensionally correct and consistent with the formal requirements of probability and information theories.

As with so many other vexing problems, that of how to introduce stocks into ascendency appears simple in hindsight. We recall that information is reckoned as the difference between the indeterminacy of the least-ordered a priori system configuration and that inherent in the actual (a posteriori) frequency distributions. In chapter 4 we estimated the a priori probability of flow from i to j to be the product of the fractions of total system activity comprised by i and j, respectively. The a posteriori probability was estimated as the observed flow from i to j divided by the TST.

It happens that we could just as well have used the biomasses to estimate the a priori probabilities. That is, the probability of a particle's leaving i can be estimated as the fraction of total system biomass that consists of species i. The same fraction calculated for species j estimates the probability that a particle enters that compartment. Hence, the a priori joint probability that a particle both leaves i and enters j is estimated by the product of the respective biomass fractions.

Using our previous example of zooplankton grazing on phytoplankton in Chesapeake Bay (see figure 4.2), we see that phytoplankton (compartment #1) comprise 31% of the entire living biomass in the Chesapeake ecosystem and zooplankton (#8) constitute 1.8% of the same total. Hence, based on a knowledge of biomass level alone, we can estimate a priori that 0.0056 of the total activity (TST) in the ecosystem will appear as flow from 1 to 8. Earlier, we saw how 0.0090 of the TST (the a posteriori probability) actually consists of grazing by zooplankton on phytoplankton. Therefore, to measure the change in probability assignment (information) we subtract the logarithm of the a priori estimate from that of the flow fraction and weight this difference by the a posteriori probability. This is done for each of the observed flows in the system, and the sum of all such terms is called the Kullback-Leibler cross-information. (The average mutual information is a special case of this more general cross-information; see Ulanowicz and Abarca [forthcoming] for details.) When the cross-information is scaled by the TST, the result can be used as an expanded system ascendency.

It can be demonstrated that the cross-information just described is always greater than or equal to the average mutual information calculated earlier using only flows. The original ascendency—i.e., the scaled AMI—henceforth will be termed the flow ascendency. In other words, the new system ascendency serves as an upper bound on the flow ascendency. The two are equal if and only if the magnitude of each flow in the system is proportional to the product of its donor and receiver biomasses. This condition is identical to the law of mass action in classical chemical kinetics. For this reason we can describe any mismatch between biomass distributions and the ensuing flows among them as representing dynamics that are "skewed" away from mass action kinetics. Increasing ascendency is abetted, therefore, by progressively skewed kinetics.

It would be serendipitous if the cross-information between stocks and flows were bounded from above by the indeterminacy (diversity) of the biomass distributions. Such a condition would imply that biomass diversity provides an upper limit on ecosystem growth and development. Ever since Robert May (1973) demonstrated that biodiversity does not necessarily abet the stability of linear systems, the worldwide movement to preserve biodiversity has had to wage its fight without adequate support from theory. A theoretical result that supports the conservation of biodiversity would be an achievement with enormous biopolitical ramifications. Alas, the cross-information possesses no least upper bound; hence, it is not possible to define a meaningful overhead, as had been done with the flow ascendency (Ulanowicz and Norden 1990). For most purposes we are forced to keep with the analysis of flow ascendency and its accompanying overhead terms. It should be noted, however, that the amount by which the system ascendency exceeds the flow ascendency can be used to gauge the degree to which the system dynamics are skewed away from mass kinetics.

If the revised system ascendency seems more ill-behaved than the flow ascendency, why consider it at all? Its advantage lies in bringing several important ecosystem behaviors under the purview of the principle of increasing ascendency. Significantly, we now may calculate how ascendency responds to changes in constituent biomass stocks. It turns out that increasing the stock of a compartment whose throughput time is longer than that of the system as a whole contributes to the ascendency. Conversely, ascendency is decremented whenever an addition is made to a species whose turnover time is faster than that defined by the aggregate system (i.e., that given by the quotient of the total biomass divided by TST). As a consequence, *increasing system ascendency now implies selection in favor of taxa having longer retention times.* Slow nutrient exchange is one of the system attributes that E. P. Odum (1969) cites as indicating system maturity (see table 4.1). Allometry tells us that longer throughput times are strongly associated with larger body sizes (Platt 1985). Hence, a tie between an organism characteristic and the ascendency finally has been forged. Higher ascendencies are now linked to bigger organisms that accumulate and retain resources.

The common practice in ecosystems research is to use chemical elements to quantify the resources that play the major roles in ecosystem dynamics. Thus, we speak of nitrogen as pivotal to the dynamics of marine systems (Ryther and Dunstan 1971) and phosphorus as the linchpin to any description of freshwater ecosystems (Edmondson 1970). As a consequence, most investigators choose nitrogen to quantify trophic flows in marine ecosystems, whereas they focus more upon phosphorus to build networks of freshwater systems (Christian et al. 1992).

The question arises, however, whether any single currency is best for portraying all critical exchanges among all elements of an ecosystem. For example, although nitrogen is most likely the element in shortest supply to marine phytoplankton, is it also the one of greatest (direct) importance to higher marine predators? No one yet knows for sure. We may still be ignoring important keys to the dynamics of higher trophic levels by quantifying them only in terms of nitrogen.

To avoid missing important information, investigators should, whenever feasible, use several relevant currencies to express the same trophic exchanges in an ecosystem. For example, three parallel networks might represent the same set of ecosystem interactions—the first using carbon (often a surrogate for energy) as the medium of exchange, the second employing nitrogen, and the third, phosphorus. We could then compare the magnitudes of each chemical element present in any particular transfer to determine the relative importance of each medium. Perhaps the most powerful way to determine which element is most important to each species is to employ information theory. In order to make such an analysis we begin by defining T_{ijl} as the amount of element l that is transferred from donor species i to receptor population j. We then use the same formula developed to express the temporal ascendency (Pahl-Wostl 1992), to define an analogous measure that pertains to simultaneous networks of several media (Ulanowicz and Abarca forthcoming).

Once again, we ask how the overall ascendency thus defined is affected by minute changes in individual stocks and flows of the several elements. It turns out that *for any given taxon, system ascendency benefits most by an increase in the element that is retained the longest by that compartment.* This result accords with Herbert Simon's observation (1956) that the commonness of resources varies inversely as the size of the inventory (as measured in time) that populations retain (Ahl and Allen forthcoming). For example, humans retain enough oxygen for three minutes; sufficient water for two days; and a store of food to last two weeks. Furthermore, it can be demonstrated that the element retained longest by a population is none other than the one made available to that compartment in least relative proportion (Ulanowicz and Baird forth-coming). Long ago, Justus von Liebig reasoned that the growth of a plant is dependent on the amount of that food stuff which is presented to it in minimum quantity (1854). His "law of the minimum" was informed by earlier observations in chemistry that reactions consume compounds in fixed proportions, and that the reactant initially present in least propor-tion is the one that eventually limits the course of the reaction. Liebig's "law" has reigned virtually unchanged in ecology since the latter began. Now, however, we can regard it as a deductive consequence of the more overarching principle of increasing ecosystem ascendency.

In order to apply this new analysis, Daniel Baird and I (Ulanowicz and Baird forthcoming) have estimated seasonal flow networks of car-bon, nitrogen, and phosphorus in the ecosystem of the mesohaline stretch of Chesapeake Bay. Because this system is intermediate between freshwater and marine ecosystems, there remains some ambiguity as to whether nitrogen or phosphorus controls system dynamics. Analysis of the sensitivity of ascendency to changes in stocks of elements in various taxa reveals that nutrient controls are indeed mixed. Most of the lower-trophic-level producers and herbivores have access to nitrogen in least relative proportion. For nekton (mobile fishes) and bacteria, however, phosphorus is proportionately least available. As for the fishes, such rel-ative scarcity reflects their heavy need for phosphorus to build skeletal tissue; bacteria, in their turn, require copious amounts of adenosine tri- and diphosphate (ATP and ADP) to serve as energy carriers to facilitate their prodigious catabolic activities.

5.4   A Consilience of Inductions

By now it should be apparent that information theory, in the particular guise of ascendency and related variables, has exciting potential for quantifying and elucidating ecosystem dynamics. Its relevance in iden-tifying nutrient controls in ecosystems is but a foretaste of such applica-tion, the depth of which I hope to plumb in chapter 7. First, however, I will pause to consider the breadth or scope of the ascendency notion. In the next chapter I describe several other theories about the development and evolution of living systems and attempt to demonstrate how much they have in common with the principle of increasing ascendency. The hope is that ascendency, in the end, will prove itself robust by a "con-silience of inductions" (fecundity of deductions) as proposed by William Whewell (1847:65).

6

OTHER MEMBERS OF THE ELEPHANT

I have never maintained that the views surrounding ascendency theory are radically original or unconnected to other current descriptions of development and evolution in living systems. To formulate a theory having no foundations in earlier and contemporary scientific and philosophical thought would be wantonly irresponsible. Thomas Kuhn (1970) warns us that the threshold to the emergence of a new paradigm is signaled by *widespread* ferment and discontent with the status quo. Many are swept up in the fervor to revise the prevailing description of nature. Like the cliché of blind persons describing the separate parts of an elephant that each feels, a great variety of ideas pours forth. There usually is some veracity in each version of how events transpire. At this juncture, then, it would be both premature and counterproductive to claim that one particular theory is correct and another is categorically wrong. Rather, during this inchoate stage we learn more collectively by trying to probe each other's perspectives to ascertain just how the emphases of our colleagues differ from our own.

6.1   Thermodynamic Conservation

In chapter 2 I attempted to outline some of the conceptual antecedents to ascendency. In brief, ascendency emerged largely from the thermodynamic perspective, which from its very inception has provided a counterpoint to Newtonian mechanistic thought. Since the early eighteenth century those who accept the Newtonian postulates have been the overwhelming majority, whose consensus determines what is favored in science. Even in the face of such solidarity, however, there remains a dialectical thrust to thermodynamics that will not be papered over, not even by acclaimed reconciliations with Newtonian mechanics, such as that purported by statistical thermodynamics.

Thermodynamics *emphasizes* the fundamental tension between order and disorder, between ordering processes (work) and disordering trends (increasing entropy). This tension found explicit quantitative expression in the "work functions" of Hermann von Helmholz and Josiah Gibbs. Energy can *never* be translated entirely into order (through work), but always entails some dissipation. Helmholz's idea was that the work that can be accomplished by energy in a given state can be expressed as the difference between the capacity inherent in that energy and the waste that necessarily accompanies its expenditure. The same concept is basic to the definition of ascendency. That is, the ordered functioning of an ecosystem (ascendency) constitutes the difference between the capacity for development in its myriad trophic connections and the encumbrance represented by dysfunctional or redundant processes (overhead) that can never be obviated. In both a formal and a conceptual sense, ascendency was patterned intentionally to become an ecological "work function"— a scion of Helmholz and Gibbs.

Of course, the classical thermodynamics of Helmholz and Gibbs pertains only to equilibrium systems, from which living systems are categorically excluded. As a flow variable or power function, ascendency belongs to the realm of irreversible, nonequilibrium thermodynamics. Traditionally, flow variables have taken a back seat to the concepts of state variables and thermodynamic forces. But we have seen that the definition of thermodynamic "forces" seems in hindsight to have been an ill-considered attempt to recast thermodynamic notions back into Newtonian terms (albeit counter to Newton's own conception of "force").

In spite of such drawbacks and inconsistencies, the irreversible thermodynamics of Lars Onsager did, at least, address collections of processes as a single whole. This was made apparent through the early work of Ilya Prigogine (1945), who used Onsager's fluxes and forces to recast the Le Châtelier–Braun principle. In Prigogine's narrative, the forces and flows occurring in any system near to equilibrium would mutually adjust to yield the configuration that produces the least amount of entropy under the existing external constraints. Whatever its limitations may have been, Prigogine's formulation provided legitimacy for the idea that nonliving systems may act in a particular direction without exhibiting any of the "purposefulness" so disdained by the critics of "teleology in biology."

Prigogine was well aware of the stringent limitations imposed by his requirement that the system remain very near equilibrium. For a while, he labored in search of a nonlinear description of how systems behave further from equilibrium (Glansdorff and Prigogine 1970). Eventually, he lost interest in purely macroscopic descriptions of systems to develop instead a theory of how abrupt changes in the configurations of metastable systems are effected by microscopic fluctuations. His model of "order through fluctuations" stands as one of the more potent counterexamples to Newtonian determinism to have appeared this century. Nonetheless, it is clear to anyone reading Prigogine that, like most biologists, he restricts causal agency to the microscopic realm. On one of the rare occasions I have been able to speak directly with him I asked why he had abandoned macroscopic description to focus upon microscopic causes. He replied that issues regarding purely macroscopic behavior seemed "trivial" to him: everything of interest was happening at microscopic scales. It was apparent, at least then (March 1985), that Prigogine had not adopted Popper's model of a universe that is causally open at all scales.

6.2   Expansionism

What biologists object to in Prigogine's principle is the conservative behavior it forces upon systems. That is, such communities always must act in a way that *minimizes* their activities. This is contrary to Darwinian logic, which awards the competitive advantage to systems that elevate their activity level to appropriate those resources that a Prigoginian system was conserving. When, long ago, I once asked Howard Odum whether ecosystems follow Prigogine's principle, he replied, in effect: "Any system that acts only to minimize its entropy production has a death-wish!"

The ecological world of "red tooth and claw" hardly seems one dominated by conservative behavior. In fact, it was Alfred J. Lotka, a Baltimore actuary, who proposed that those organisms that achieved a high output relative to their body size should win out in the competitive struggle for existence (Lotka 1922). Howard Odum has been Lotka's champion in latter-day ecology, providing copious examples of the "Lotka maximum power principle" at work imparting structure to dynamical systems (H. T. Odum 1971). He translates Lotka to read that systems perform at that efficiency which results in maximum power output. Such efficiency is always less than the highest possible under the first law (H. T. Odum and Pinkerton 1955).

Although Odum uses a thermodynamic argument, he eschews the language of that field, almost never mentioning entropy or its production. Others are less reticent. Rod Swenson (1989a,b), for example, believes that the overarching attractor in the universe is the tendency for all systems to create entropy at the fastest rates consistent with their external and internal constraints. One immediate problem with Swenson's principle is that it places special emphasis upon the interplay among thermodynamic forces, which, as we have seen, are ill defined to begin with and usually remain elusive. Another problem is that one's interest is very quickly distracted away from this monistic view of development, toward the question of what opposes maximally fast dissipation. The universe, after all, does not burn up in a flash; presumably, its reprieve is due to the existence of constraints. In Swenson's formulation, however, these constraints always must be considered on a particular or ad hoc basis.

Rather than treating constraints as external to the principle, one might attempt to incorporate the constraints into the defining variables. This is what Helmholz and Gibbs did with their work functions: Available work is the capacity for change that is left after the internal microscopic interferences (constraints) have been subtracted. The big problem with the Gibbs and Helmholz functions is that they can be defined only for systems at equilibrium. Robert Berton Evans (1969) was able to generalize the idea of available energy so that it pertained to nonequilibrium systems as well. His broader measure of the work that the energy in a system can accomplish is called the *exergy*. In addition, exergy serves as a measure of the distance of the system from thermodynamic equilibrium.

Using exergy to reformulate the second law does indeed internalize some constraints (Keenan 1951). But constraints are the building blocks of organization. So the second law, restated in terms of exergy, no longer implies a rush pell-mell toward stochasticity. Thus, when Eric Schneider and James Kay (1994a,b) posit that living systems respond to an imposed gradient in exergy in such a way as to reduce (dissipate) that gradient, it is the implied internal constraints that guarantee that such dissipation will not occur at the fastest rate possible. In fact, Schneider and Kay go on to claim that if the imposed gradient should be increased, the system will respond by building more structure to oppose that increment. Nevertheless, although dissipation by a living system does not occur at a maximal rate, it does proceed faster than if all life forms were absent. (See also Ulanowicz and Hannon 1987.)

Confusion may temporarily arise as we now note that Sven Erik Jorgensen and Henning Mejer (1979) contend that systems generally act to store within themselves as much exergy as possible. In other words, the same systems that are rapidly destroying exergy in the world around them are at the same time busily increasing their own stores of exergy. Although it may seem that the dissipative observations of Kay and Schneider are hopelessly at odds with the conservative hypothesis of Jorgensen and Mejer, the ostensible conflict is resolved through hierarchical reasoning. A developing system occupies the focal level of our attention. It extracts exergy from the next-higher scale at a rapid rate. Some of the available energy is used by the system to create ordered energetic structure (exergy) at the focal level; the bulk, however, is dissipated at other levels. Overall, exergy disappears. This resolution of the paradox follows closely Prigogine's explanation (1967) of how the entropy of a particular finite system can decrease while embedded within a universe of ever-increasing entropy; indeed, it generalizes the principle.

6.3   Thermodynamic Dialectics

Out of the thermodynamic issues surrounding development, two dialectics emerge. The first is between those who see a world that behaves conservatively by minimizing its activity and retaining resources (Prigogine and Jorgensen), and those, on the other hand, who see one populated by highly expansive and dissipative elements (Lotka, Odum, Swenson, Kay, Schneider). The question is whether or not this clash of outlooks mirrors the existence of an actual conflict in the nature of things. Kay, Schneider, and Swenson would argue in the negative, maintaining that the second law reigns paramount and self-sufficient. Stanley Salthe (1993), Lionel Johnson (1990), and I believe that what is represented by this "clash of outlooks" is a reflection of actual dynamics.

Lionel Johnson spent much of his career collecting long-term data on the fish populations and other biotic components of high-latitude lakes. He found that when ecosystems are relatively isolated, competitive exclusion results in a relatively homogeneous system configuration that exhibits low dissipation. In contrast, physical dissipation is much higher in ecosystems that are less fully developed (or whose development is arrested due to external influences): over evolutionary time their taxonomic heterogeneity appears to increase. Johnson concluded that "ecosystem structure is a function of two antagonistic trends; one toward a symmetrical state resulting in least dissipation, and the other toward a state of maximum attainable dissipation" (1990:14). In other words, ecosystem organization is the outcome of a clash between two opposing, but mutually obligate, trends.

Population ecologists will immediately see in Johnson's remarks some resemblance to the *r* vs. *K* dichotomy in ecosystem development, formulated long ago by Robert MacArthur and Edward Wilson (1967). The letters r and K refer to the two parameters in the logistic model for population growth. When resources are abundant, conditions favor the proliferation of species that grow (and dissipate) at high rates (r). Gradually, resources are depleted by such growth, and the advantage in competition shifts to the "K-strategists," who persist in relative harmony with their environment and coinhabitants and sequester resources at minimal expense (dissipation). There is an analogy here, and maybe some shared causality. Readers should bear in mind, however, that the r-K narrative pertains only to single populations. Johnson's observations, by contrast, encompass entire ensembles of interacting species and their environment, i.e., *ecosystems*.

How, then, to enfold the Prigoginian emphasis on conservation and Lotka's focus on expansion into a single, overarching, *quantitative* statement was the challenge that first motivated the formulation of system ascendency. A tension between these opposing trends has been built into the ascendency. When external conditions become extreme, it is possible that one of the conflicting trends will prevail. For example, whenever external resources become quite abundant, the activities of those elements that can appropriate (and dissipate) them as quickly as possible dominate the system. The rise in ascendency during flush conditions is due almost entirely to increasing total system throughput. At the other extreme, when external resources are meager, those taxa persist that can, in cooperation with like-behaving species, sequester resources with a minimum of dissipative overhead. Under sparse conditions, which usually prevail later in the course of system development, a rise in ascendency almost always is due to an increment in the mutual information of the flow structure.

The second dialectic in the description of system development occurs between those who emphasize contents and those who emphasize transformations—i.e., stocks or flows. Classical thermodynamics dwells upon state variables, those attributes that describe a system at equilibrium. Conceivably, transformations can take place in a system at equilibrium; such flows, however, are always balanced, and their associated dissipations are negligible. For these reasons flows remain in the background of classical thermodynamic descriptions, which focus by default upon the contents of a system. Should a system undergo some transformation and return to its original configuration, the change in each of its state variables would be zero by definition. The amount of activity (flow) associated with the excursion is not unique but rather depends upon the exact nature (pathway) of the processes that comprise the cycle. Once again, we note how the conventional outlook emphasizes conservative state variables, such as pressure, temperature, or chemical composition, over nonconservative process variables, such as heat flow, dissipation, or material throughput.

In formulating the concept of exergy, Evans (1969) sought to deal with this bias by extending the notion of a state variable as far as possible to nonequilibrium situations. His exergy connotes the energy that *potentially* could appear as work. Exergy thus measures a content or stock. The original definition of ascendency, by contrast, was based entirely upon flows or transitions. Only recently have I been able to integrate stocks into an amended definition (see chapter 5). Thus, the contrast between Jorgensen's principle of maximal exergy storage and the ascendency hypothesis consists in the dialectic between a description of component stocks and a narrative of intercompartmental flows.

Pertinent to this comparison is Bernardus Dominicus Hubertus Tellegen's theorem from network thermodynamics (Mickulecky 1985). The goal in network thermodynamics is to describe a system by linking a collection of "through" variables (flows) between nodes to a conjugate set of "across" variables that take the form of differences in some potential that exists between the nodes. Tellegen showed that corresponding to any such description there exists an equivalent "dual" formulation wherein the "through" variables of the former correspond to the "across" variables in the dual, and vice versa. A behavior can be described using either the original problem statement or its dual.

Network thermodynamics, despite its name, is primarily a mechanical description of nature. One does not expect Tellegen's theorem to apply to thermodynamics at large. Yet it would seem that some connections must exist between state descriptions and flow narratives. That a system might increase in both exergy content and ascendency during the course of its development is not only conceivable, but likely. As Jorgensen (1992) suggests, it is highly plausible that we are creating pluralistic descriptions of the same phenomena. Which description one prefers turns upon one's own predilections and what information is at hand.

As mentioned earlier, it is usually far easier to take data on stocks than to measure flows. Such convenience, when amplified by the historical bias that favors state over process variables, would seem to favor heavily the principle of increasing exergy. It happens, however, that estimating the exergies of living organisms is no simple matter. For in order to calculate the exergy of a component, it is necessary first to estimate its entropy—and as we have seen earlier, determining the entropy of a living system is a highly problematical exercise (Christensen 1994). By comparison, however, obtaining the ascendency of a trophic exchange network is conceptually straightforward, albeit laborious in execution.

The possibility remains that system ascendency and exergy are related in a manner just short of equivalency. For example, abiotic elements are part of each ecosystem, and it is no great problem to estimate the exergies of these physical compartments. If it were possible somehow to gauge how much the exergies of the living components exceed those of their abiotic cohorts, then the exergy values of all components could be determined. Toward this end, we recall that when the accounting medium is energy, the ascendency becomes a work function. That is, to obtain the contribution of each energy flow to the ascendency of a network, one multiples its magnitude by a "quality factor." This factor represents the concentration of lower-quality energy from the donor into the potential for doing high-quality work by the receptor. That is, within developing systems, flows are usually "uphill" toward more concentrated forms, presumably maintained by informational constraints (Ulanowicz 1972). (Bear in mind that a preponderance of the exergy from the donor is dissipated during the process, so that neither the Kay-Schneider observation nor the second law is contradicted by this scenario.) We note also in passing that the "uphill" nature of these flows means they are not spontaneous, but rather the result of an agency (propensity), in the same sense that Newton used to define his forces.

We immediately recognize the logarithmic terms in the energy-based ascendency as the quality factors that can be employed to estimate the differences in exergy between connected nodes. Hence, starting with the known exergies of the abiotic components, we may be able to invoke these factors to build up the implied exergies of the higher trophic elements. One problem with this strategy is that there usually are more flow connections than nodes in a system, so that the exergies of the living components would be overdetermined by this method. In practice, however, overdetermined systems can readily be handled by optimization techniques such as singular value decomposition (Golub and Van Loan 1983), to obtain the best concordance between the exergetic and ascendency formulations.

6.4　Peripheral Associations

The principle of increasing ascendency and the two exergetic hypotheses all appear to subsume at least two opposing trends, such as those noted by Prigogine and Odum, respectively. A number of other monistic directions have been ascribed to developing systems that at times overlap or parallel these compound narratives. In this regard, I have already mentioned Swenson's opinion that systems develop so as to maximize their production of entropy—a direction that often parallels an increasing ascendency under many (but certainly not all) circumstances. Likewise, Daniel Brooks and Edward Wiley (1986) identify "evolution as entropy." By "entropy," they mean the Shannon (1948) diversity of species within a particular evolutionary tree (clade). Simply put, Wiley and Brooks observe that between large-epochal catastrophes, the number of species in any cladogenetic line tends to increase. Although they have been strongly criticized for their lack of thermodynamic rigor (Morowitz 1986), their observations nonetheless are congruent with the increase in development capacity that occurs during the intermediate stages of system development.

Looking more in the direction of increasing order, Bernard Patten and Masahiko Higashi (1991) have emphasized that much of the causality behind changes in ecosystems is expressed via indirect pathways. More importantly, they cite the tendency for indirect influences that began as antagonistic or indifferent in character gradually to become mutualistic in nature. Along the same lines, we have already remarked how autocatalysis often occurs over indirect pathways, and how such positive interactions gradually tend to displace other types of interplay during the course of building ascendency.

Numerous investigators have referred to the tendency of ecosystems to capture and store material and energy. Edward Cheslak and Vincent Lamarra (1981), for example, see the average time that energy remains in a system as the key index to ecosystem maturity. In much the same vein, Bruce Hannon (1979) echoes Ramon Margalef's earlier (1968) suggestion that during development the amount of metabolic loss per unit of stored biomass declines. Both observations accord with my earlier suggestion that the drift toward increased ascendency favors those compartments that retain medium longer than average. It is not only storage but also cycling among components that can increase the time a typical quantum of medium remains within the system (Finn 1976). As we have noted, increased cycling contributes to a higher TST.

6.5   Laws or Orientations?

Many of the foregoing descriptions of system development have been stated as "variational principles"—that is, some quantity is being either maximized or minimized during the course of system development. (Indeed, my early statements of the ascendency hypothesis were couched in just this way.) The danger in the variational approach to directional change is *not* that optimization implies teleological behavior: we saw how Hamilton's principle for moving bodies and Prigogine's law for near-equilibrium systems provide convenient examples of variational behavior devoid of teleological implications. Ironically, the problem is that variational behavior imputes a far too mechanical nature to living systems. To impose variational "laws" upon living systems would be to proscribe too much of their freedom and ambiguity. In order to distinguish the directional drift of developing systems from any form of mechanical determinism, Hartmut Bossel (1987) suggests that we refrain from speaking of "goal functions" at work and speak instead of "orientating functions" (cf. Depew and Weber 1994). For much the same reason, the tendency of system ascendency to increase is now being stated without placing any qualifiers on how such increase should occur. It would seem in hindsight that most of the variational statements discussed in this chapter were better stated in less-deterministic terms.

6.6 Complexity Theory

Having discussed the common features among those theories of development that derive from thermodynamics, I now consider how the ascendency principle stands in relation to other hypotheses that are informational in nature, but not thermodynamic in origin. Perhaps the most celebrated recent outlook on the sources for order in living systems is known as "complexity" theory (Waldrop 1992; Kauffman 1991). Complexity theory has become the centerpiece of the Santa Fe Institute (SFI), whose investigators have provided an impressive series of examples that employ cellular automata, neural networks, and genetic algorithms to create patterns akin to those that appear in developing organisms and ecosystems. One of the most intriguing branches of the complexity school is the burgeoning quest for "artificial life" (AL), whereby a class of cellular automata is created that gives rise to dynamic patterns that resemble self-organizing entities (e.g., Langton 1992; Forrest 1991).

These ingenious constructs provide significant insights into the mechanical aspects of the life process. It should not surprise anyone who has read this far, however, that I see strict limits to just how far such mechanical approaches can carry us toward an adequate understanding of living systems. Save for some mild castigation from within SFI itself (Casti 1989, 1992), complexity theory represents "business as usual." In complexity theory, the reductionist assumption still prevails that one can achieve an understanding of living behavior in solely algorithmic (mechanical) terms—the only hitch being that the mechanics of life are a little more complicated than usual. This attitude is explicit in models built on cellular *automata*, and it is only partially hidden in the neo-Darwinian genetic *algorithms* that juxtapose mechanical behavior with unconditional stochasticity. Neither treatment approximates conditions in Popper's open universe, nor provides an avenue to develop a calculus of conditional probabilities.

One irony about complexity theory is that its practitioners seem less occupied with complexity per se than with limits to complexity. Christopher Langton (1992), for example, presents a model suggesting that life exists at the "edge of chaos." He identifies four classes of cellular automata, the behavior of which grades from stable fixation (Type I behavior) to stable oscillations (Type II) and thence into chaotic excursions (Type III). Within the transition from oscillatory to chaotic behavior, however, he demonstrates the existence of a narrow range of new, Type IV dynamics, which he likens to a "Lambda" type phase transition in physics. In Type IV automata, "identifiable" clusters of cells appear that maintain their integrity, move about, and interact with each other in rudimentary fashion. Because these more complicated "lifelike" forms exist in only the very narrow range of parameter space that separates the mundane from chaotic behaviors, Langton bids us infer that life itself is poised on the edge of chaos.

Another SFI investigator who has examined the boundary of chaos in dynamical systems is Stuart Kauffman (1991). Kauffman departs from the traditional way of representing systems of genetic transitions (usually as cellular automata) and portrays them instead as *networks* of transitions among genetic configurations. He then surveys the number of connections per node found in networks that yield realistic behaviors. Among them we do not find arbitrarily complicated systems with, say, seven connections per element. Rather, the behaviors of Boolean networks of genetic interactions remain chaotic until the number of links per element drops to about three—whereupon the networks exhibit unexpected, spontaneous collective order. Interestingly enough, Stuart Pimm (1982) estimated that the number of corrections per node in his collection of ecosystem food webs averages about 3.1. This coincidence only deepens when we see that Jorge Wagensberg, Ambrosio Garcia, and Ricardo Sole cite a "magic value of about 3 bits per emitter [as] an actual upper limit to connectivity in real stationary ecosystems" (1990:739).

Because Kauffman has represented gene dynamics in terms of networks, his work serves as a point of contact between complexity theory, evolution, and ascendency. It can be demonstrated that the conditional indeterminacy of a network of ecosystem flows (the Shannon diversity of flows minus the average mutual information—see chapter 4) is a convenient measure of the effective number of connections per node (Ulanowicz forthcoming). Information theory can also be used to quantify the intensity of interaction between nodes. Coupling between two compartments is considered maximal (= 1) if the connection between them is the only output of the donor and the sole input to the receptor. The magnitude of interaction becomes a progressively smaller fraction as interactions with other parties grow stronger. Using dimensional analysis to compare the relative energies of cohesion (intensity of interaction) with those of dispersion (number of connections per node) leads to the calculation of a greatest upper bound on the number of connections per node in stable systems: that limit is $e^{3/e}$, or about 3.015.

It would appear from this example that some of the dynamics of living systems become accessible via the "calculus of conditional probabilities," which provides an understanding not afforded by the mechanical approach. In fact, the artificiality imposed by a strictly mechanical portrayal can induce spurious conclusions. For example, when we calculate the limit to stability as outlined in the last paragraph, we discover that data on observed ecosystem networks do not crowd up on the "edge of chaos," as Langton's automata would lead us to believe. Rather, the available data reveal that most systems are removed both from the edge of chaos and from the "brittleness" of one-on-one obligate connectivity (Ulanowicz forthcoming). That is, they tend to inhabit a dynamical "window of vitality" that can be characterized neither as chaotic nor as frozen.

To depict this result more clearly, I have plotted how two types of connectance are related to each other in thirty-eight different ecosystem trophic flow networks as estimated by seventeen separate investigators (figure 6.1). Measured along the horizontal axis is the "topological connectance" of each system. This index is the log-mean average of the number of links both into and out of each compartment of a particular network, irrespective of the relative magnitudes of those connections. The vertical axis represents the "effective connectance" of the same networks, which is calculated using the same formula as for the topological connectance, except that in the effective connectance the links are weighted by their relative strengths.

Because of the weighting and averaging scheme used, it is algebraically impossible for real networks to plot into the area above the 45° line in figure 6.1, labeled "infeasible." The single curved line represents the information-theoretic homolog to the May-Wigner stability criterion (May 1973); that is, only systems with combinations of connectances that plot

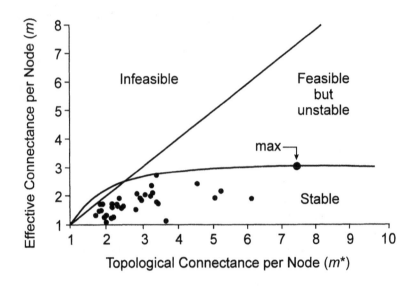

FIGURE 6.1.
Domains of stability that can be occupied by a system flow network, according to the exact nature of its connectivity. Topological connectance is a log-weighted average of the number of connections per node. Effective connectance is calculated in the same manner, except that each flow is first weighted by its own magnitude. For algebraic reasons, all systems plot below the radial line. Stable systems should occupy the region below the curve. Dots indicate where the estimates for thirty-eight real networks fall.

below this demarcation have any real chance of persisting. This curve also represents Langton's "edge of chaos" in that points above this border are either unstable or infeasible. The curve possesses an indistinct maximum at about 3.01 connections per node, as indicated. The horizontal axis itself represents all combinations for which the effective connectance equals one—i.e., each node of any system falling on this line would have but a single link into and out of it; it would be minimally connected, maximally obligate (mechanical), and very brittle, or vulnerable to external perturbations. The area sandwiched between the axis and the curve represents the "window of vitality"; all thirty-eight systems plot inside this region, and they do not crowd upon the "edge of chaos."

Popper says that forces are particular, degenerate examples of propensities (see section 3.1). Viewed in this light, Langton's mechanical treatment has artificially and artifactually compressed the "window of vitality" into the "edge of chaos." That his approach might have benefited from the inclusion of probabilistic considerations is hinted at by Langton himself (1992) when he points out that the mutual information that one cellular configuration bears to its next iteration becomes maximal in the narrow Type IV region. It is altogether possible that with somewhat fewer automata and a little more chance and information theory the models of AL could become less artificial.

6.7   Chaos

In discussing complexity theory I have made several references to "chaos" or "chaotic behavior." Since chaos theory occupies the attention of so many these days, I should discuss the topic at least in passing. Before I do so, it should be noted that chaos is but one of several types of dynamics that have elicited widespread interest during the past two decades; among the others are catastrophe theory and fractal dynamics. The vehicle I have chosen to describe ecosystem development—information theory applied to networks of transitions—is quite robust with respect to underlying dynamics. It has to be, in fact, if it is to be usable even when there are causal gaps in a system's dynamics. Hence, ascendency and related variables should be appropriate to systems governed by any category of dynamics.

It may happen, of course, that invoking ascendency to treat deterministic situations will prove uninteresting. For, as Salthe (1993) tells us, mechanical dynamics do not represent essential change very well. Deterministic models, comprised of interactions represented as "fixed functional forms," exhibit behavior that can be plotted into a limited (but not necessarily contiguous) region of dynamical phase space. Over time, the behavior of a nonchaotic system comes to occupy a diminishing portion of this subspace. Chaotic systems behave differently: over a sufficiently long interval, a chaotic system will continue to visit all its allotted phase space. Be it chaotic or otherwise, however, if we sample a deterministic system long enough, any ensuing probabilities for transitions will converge to static distributions. This is precisely what makes deterministic systems uninteresting to those seeking to understand true change. Because information theory, by definition, quantifies changes in probability assignments, it is admirably suited to the mensuration of true change, regardless of underlying dynamics (Ulanowicz 1993).

As to chaos theory itself—virtually everyone agrees that its implications for prediction in science are revolutionary. There is no consensus, however, on exactly how it will affect scientific thought. Do the insights provided by chaos theory expand the realm of deterministic behavior, or delimit its boundaries? It would seem from some popular accounts that chaos is being invoked to preserve the notion of determinism and reductionism. Complicated behavior, like turbulent fluid flow, now can be considered deterministic in principle: if only one knew the starting and boundary conditions with sufficient precision . . . , etc. Relevant to boundary conditions is the much-heralded "butterfly effect"—the notion that the flutter of a butterfly's wings in Indonesia could precipitate a hurricane that eventually crashes into Mexico. As with Prigogine, popular imagination seems preoccupied with the idea that the causes of events always lie at smaller scales. It seems that for many, determinism and reductionism will reign, come what may!

Others, however, see in chaos theory reasons to put strict reductionistic determinism behind us and get on with building a new picture of how the world works (Goerner 1993). It is fantasy to think that we can measure conditions to arbitrary precision. We can never isolate systems entirely from interfering events, and chaos theory says it takes only the slightest interference to set the system on a radically new trajectory. What chaos theory *really* seems to be showing us is the futility of the dream of precise prediction (Ulanowicz 1979). Of course, a limited prediction ability can be deduced from the shape of the "strange attractor" that maintains a chaotic system in its proper subspace. But is it necessary to postulate that it is mechanical constraints that define this subspace? Quantitatively, the same degree of prediction could be achieved by calculating the conditional probabilities and information measures introduced here. Some will object that the probabilistic approach does not reveal causality in the mechanistic sense—but if it is mostly formal and final agencies that lend cohesion to the dynamics, then little has been lost.

As for the butterfly effect, it is always presented looking forward in time. Look at chaotic systems in reverse, however, and one is inclined to draw very different conclusions. In hindcasting it becomes progressively more difficult to trace the antecedents of a chaotic system. Chaotic systems "forget" their pasts too easily. There is simply no hope of studying the Mexican hurricane and identifying the culprit butterfly. Mechanists will rejoin that just because one cannot precisely specify the initiating cause this does not nullify the assumption that an initiating mechanism must have existed. Tenacity to this belief belies an attitude that the macroscopic is trivial (to use Prigogine's words)—that causal agency resides only in the microscopic.

There is an egalitarian streak that permeates most examples of "order through fluctuations": it is that all fluctuations have an equal chance to precipitate macroscopic reordering. The implicit assumption is that a macroscopic system does not remain immune to most types of perturbations. However, even in metastable systems, only a subclass of possible perturbations stands any chance of reordering the macroscopic configurations. It is not simply a matter of *any* arbitrary perturbation arriving on the scene to punch the system into a new configuration. Exactly which subclass remains capable of initiating change is set by the macroscopic state of the system. It is a subtle causal exchange across hierarchical scales. The larger system delimits the nature of those perturbations to which it is vulnerable, and its original state constrains the possibilities for its subsequent reconfiguration. The butterfly effect is so much blue smoke and mirrors distracting us from this elegant dance of causality that transverses hierarchical levels.

This pattern of response to unsettling new observations is not new. The Newtonian worldview remains tenaciously entrenched among the scientific body politic. Every time a new observation or idea appears to set the accepted picture akilter, ways immediately are sought to isolate or obviate the offending notions: Statistical mechanics became the antidote to thermodynamics. The "grand synthesis" banished biological indeterminacy to the microscopic realm of atoms and genes. Now chaos theory is being slanted to exaggerate the role of reductionistic determinism in nature.

If these philosophical considerations seem tedious to some, it is probably because they deem them to be of little practical consequence. It is to this criticism that I now turn my attention, as I pursue the ramifications of an "open universe" and ascendency upon ecology, science, and human attitudes in general.

7

PRACTICAL APPLICATIONS

Thus far, we have discussed mostly concepts surrounding the focal notion of ascendency. But ascendency is more than a vague notion. It derives from a very concrete mathematical formula that can be applied to any ecosystem whose connections can be identified and quantified. Once we know which system component affects which others and by how much, we can *measure* ascendency. The quantitative answer to "by how much?" need not be expressed on an absolute scale—it is necessary only that all measurements have the same dimensions. (See Hannon, Costanza, and Ulanowicz 1991 for ways to unify disparate dimensions.) Given a suite of such numbers, it is a routine matter to estimate the system ascendency, capacity, and overhead components, which then provide a profile of the system's "instantaneous" developmental status. Of course, these indices change as the system changes. Noting how they are altered and by what degrees allows us to draw some explicit conclusions about the changes in system status.

7.1   Quantifying Ecosystem Status

Possibly the most pertinent use of the ascendency measures is to quantify the effects of perturbations on ecosystems. A whole-system index that gauges how much a system has changed in response to a particular disturbance should be a valuable tool for evaluating ecological mitigation projects. Unfortunately, sufficient data to describe the full suite of trophic interactions as they occur in an ecosystem both before and after perturbation are exceedingly rare. The unpublished data of M. Homer,

W. M. Kemp, and H. McKellar (1976), however, on the flux of carbon through the ecosystems found in two tidal marsh creeks off Crystal River, Florida, comprise a notable exception (see also Ulanowicz 1984, 1986a). The first of those creeks serves as the natural benchmark. Homer, Kemp, and McKellar have identified the major taxonomic groups that make up this unperturbed example, including both lower trophic components and various fishes. The flows between them were estimated in terms of mg carbon/m^2/d and are depicted as arrows in figure 7.1. The second nearby creek was similar to the first in all ways, save one: it was subjected continuously to heated effluent from a nearby nuclear power station. The temperature difference between the two creeks averaged 6°C over the year.

These comparative data should allow us to test the predictive adequacy of ascendency measures. If we expect ascendency to increase under relatively unimpeded growth and development, then a significant perturbation should have the opposite effect on the information indices. That is, we anticipate that the total activity (total system throughput) and the development capacity of a stressed system would both be negatively affected. We note that the total system throughput does in fact fall off by almost 20% in the heated ecosystem. Because all information indices are modulated by this drop in throughput, their absolute magnitudes are all lower in the affected system.

To quantify the degree of structural difference between the two systems it is necessary to compare the *relative* magnitudes of their various information indices. The diversity of flows, for example, can be measured by the quotient of the system capacity divided by the total throughput; it drops from 3.153 bits in the unimpacted system to 3.119 in the heated marsh. The percentage of capacity that appears as ascendency (ordered structure) falls from 40.1% to 39.8%. The fractions of overhead encumbered by dissipations and flow redundancies both rise slightly in the disturbed ecosystem, as one would expect in response to disturbance. The only change not characteristic of disturbance is a slight drop in the fraction of overhead associated with exports. Apparently, the decrease owes to the fact that respirations (dissipations) in the heated ecosystem do not decline in proportion to the decrease in total activity that warming incurs. The information that is added to the flow ascendency because biomasses are "skewed" with respect to flows is slightly greater in the control ecosystem (2.352 vs. 2.347 bits), indicating a likelihood that more organisms with slower turnover rates inhabited the unperturbed system.

With one minor exception, changes in the information indices were opposite to the normal trends in growth and development—that is, they

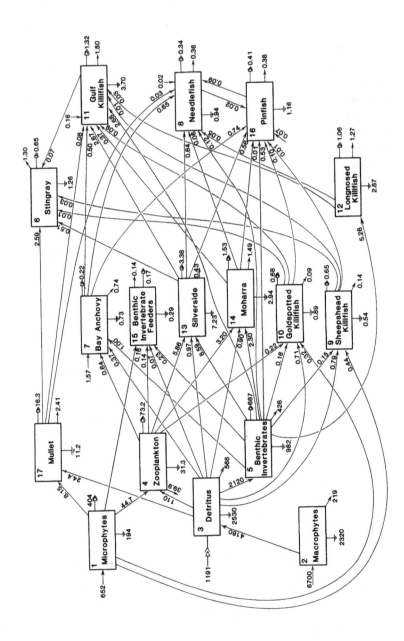

characterized a setback, or disruption to normal system functioning. That the changes were small indicates that the topology of the network did not change appreciably. (It should be noted, however, that differences in information indices represent logarithms of changes in actual probabilities, and are consequently much-attenuated indicators of change.)

With the burgeoning popularity of trophic network descriptions, it is likely that new comparisons of perturbed vs. unperturbed systems will soon appear to test further the descriptive power of ascendency and related variables. Eventually, it should become clear what changes in which information variables correspond to characteristic qualitative shifts in ecosystem behavior. Gradually, it will become possible to track quantitatively changes in the overall status of an ecosystem and to identify unequivocally when and by how much the entire system has been disturbed.

7.2   Assessing Eutrophication

Certain characteristic behaviors induce signature changes in information indices—a vivid example of which is the process of eutrophication. It has been enormously difficult to define eutrophication, and little consensus has been reached, even after several international workshops (Christian et al. 1996). The root *eutrophy* is Greek in origin and means "well-fed"; however, in its technical use the word has acquired a negative connotation: it suggests that the system in question has been overenriched. A eutrophic system is one that has received too much of a good thing. The negative consequences of overenrichment are usually manifested as the loss of important species along with their attendant system functions. In other words, nutrients tend to stimulate a system's growth, but in spite of this augmented activity, its organization is degraded.

In chapter 4, we saw how "growth" and "organization" may be quantified using flow analysis and information theory, respectively. Growth was defined as an increase in system activity or total system throughput. System organization was equated with the mutual information inherent in the trophic flow structure; thus, development is identified as any increase in the mutual information of the exchange configuration. An increase in

FIGURE 7.1.

Transfers of carbon among seventeen major components of a coastal salt marsh creek near Crystal River, Florida (Homer, Kemp, and McKellar 1976).

Flow magnitudes are in $mg/m^2/d$. "Ground" symbols represent dissipative losses, and double arrowheads depict returns to the detrital compartment (#3).

ascendency—the product of the terms for growth and development—quantifies the unitary process of how a system naturally matures. It is possible, however, for growth to occur in the absence of any development, or even in the face of waning organization. This latter circumstance describes precisely what happens during eutrophication. Therefore, an unambiguous, quantitative definition of eutrophication stated in terms of ascendency theory is "any increase in system ascendency due to a rise in total system throughput that more than compensates for a concomitant fall in the mutual information of the flow network" (Ulanowicz 1986b:80).

For an example of how we might use this definition to detect eutrophication, we can look again at the Chesapeake ecosystem network in figure 4.2. Its ascendency is 8,593,800 mg $C/m^2/y$ bits, which is the product of its total system throughput (4,116,200 mg $C/m^2/y$) times its mutual information (2.088 bits). Now let us imagine a hypothetical increase of nitrogen into the ecosystem that increases the system ascendency to 10,136,400 mg $C/m^2/y$ bits. We then ask whether eutrophication has occurred. If the increase in ascendency was because the TST rose to 4,649,724 mg $C/m^2/y$ *and* the mutual information of the flows increased to 2.180 bits, then we would say simply that the system has been enriched. If, however, the augmented ascendency occurred because the total system throughput rose to 5,449,677 mg $C/m^2/y$, while the conjugate mutual information *fell* to 1.860 bits, then we could rightly say that the system has undergone eutrophication in response to the additions.

7.3   Ecosystem Health and Integrity

Eutrophication is but one example of pathological changes in ecosystems. In managing ecosystems the goal is usually to avoid system pathologies—or, as the metaphor goes, to maintain the ecosystem in a "healthy" state. The analogy here is to the health of an organism, which basically means the absence of disease. But health, as it pertains to ecosystems at least, is usually a bit broader in meaning. A healthy or well-functioning ecosystem is also vigorous. It has an adequate potential to grow. A healthy ecosystem is also one that resists perturbation, or, if perturbed, recovers toward its unperturbed configuration. Other desiderata are that healthy systems maintain a certain balance among system components, and a sufficient diversity of taxa or complexity of functioning. Robert Costanza (1992) has suggested that these several aspects of healthy or well-functioning ecosystems be aggregated into three fundamental system attributes: (1) vigor, (2) organization, and (3) resilience.

It is obvious from the example of eutrophication that not every increase in system ascendency represents healthy change. That is, although ascendency does encapsulate several aspects of growth and development, the concept is not broad enough to incorporate all elements of ecosystem health (Ulanowicz 1992). As Wagensberg, Garcia, and Sole (1990) suggested, because there are several components to ecosystem health, we should not expect a single index to encompass the concept fully. (Wagensberg's group provided simple models of ecosystems in which pairs of system indices were used to delimit a region that characterized well-functioning systems. Their inference was that systems that plotted outside this domain were "unhealthy" in some respect.)

Let us return to Costanza's triadic view of health. It is easy to relate each property he cites to one of the ascendency variables. Vigor, for example, is well represented by the total system throughput. System organization already has been identified with the mutual information of flow networks. Finally, the conditional indeterminacy of the network (indeterminacy minus mutual information) is related to the reservoir of possible responses to novel perturbations—i.e., the ability of the system to persist and be resilient in an uncertain world. By plotting the values of these three variables for many ecosystems, it may become possible to identify a region in parameter space that characterizes healthy ecosystems.

Unfortunately, three-dimensional plots are cumbersome to draw. Furthermore, the dimensions of the two information components are incompatible with those of the system throughput. For these reasons, Michael Mageau, Robert Costanza, and I (Mageau, Costanza, and Ulanowicz 1995) have suggested a convenient mapping of the three-dimensional parameter space onto two dimensions. We note that the product of vigor (TST) times organization (mutual information) yields the system ascendency. Likewise, when the resilience (conditional indeterminacy) is scaled by the vigor (TST), the product is the system overhead. Thus, in any plot of ascendency vs. overhead, the total system throughput will appear as an implicit scaling parameter. We are currently engaged in estimating the ascendencies and overheads of a series of mesocosms that have been subjected to a gradation of controlled perturbations; it will be interesting to see if the ensuing graph of ascendency vs. overhead reveals a zone that discriminates well-functioning ecosystems from those that border on pathological behavior (see figure 7.2).

While ecosystem "health" has occupied the attention of U.S. ecologists, the corresponding watchword in Canada has been "integrity" (Edwards and Regier 1990; Westra 1994; Lemons and Westra 1995).

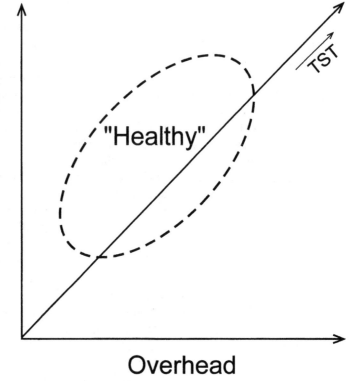

FIGURE 7.2.
Anticipated region of "healthy" ecosystems (*dotted oval*) in a plot of system ascendency vs. overhead.

Health and integrity, as applied to ecosystems, are not synonymous, however. Ecosystem health mostly addresses how well the system is functioning at the present moment. Integrity applies to a broader time horizon and includes the ability of the system to deal with unforeseen circumstances in the future (Ulanowicz 1995b). Integrity encompasses a system's entire trajectory of past and future configurations (*sensu* Holling 1992). The direction in which a system is headed (its telos) is not only an integral element of its integrity, it also can impart a legitimacy to ethical considerations of how society should interact with the system (Westra 1994). The temporal versions of system ascendency and capacity (Pahl-Wostl 1992) appear most appropriate to deal with ecosystem integrity.

7.4 Comparisons Between Ecosystems

The question arises in making comparisons between ecosystems, how meaningful are the magnitudes of information indices pertaining to different ecosystems? Obviously, if two investigators independently consider the same ecosystem, it is likely they will cast their descriptions using different sets of taxa and exchanges. (One assumes that both accounts are complete, and that decisions regarding whether or not to include spatial and temporal inhomogeneities are the same in both cases.) The magnitudes of their respective information indices obviously will not be identical. Does the subjective nature of these measures render them useless for comparative purposes?

This question can be addressed from either a qualitative or an absolute standpoint. As for qualitative changes in network information indices, they appear quite robust with respect to how one parses the system. That is, it appears that if one investigator perceives a change in any of the information indices, the chances are that the other investigator, who identifies different groupings of taxa, nonetheless will see the same qualitative change (e.g., rise or fall) in the same variable, albeit of a different magnitude. For example, if one person ascertains that the ecosystem has undergone eutrophication, then calculations by the other investigator using a different aggregation scheme should tell the same story.

Of course, we do not always compare changes in the same ecosystem. We may wish instead to compare the current conditions of two separate ecosystems. In such case we can expect to make a valid comparison only if the systems in question have been parsed in nearly identical fashion. This does not mean that the ecosystems must have exactly the same list of species or functional units: it means that they must be aggregated into roughly the same number of compartments standing in similar juxtaposition with each other, and must possess corresponding turnover times that are not radically different.

Such a comparison has in fact been made between the Chesapeake Bay (figure 7.3) and Baltic Sea (figure 7.4) ecosystems (Ulanowicz and Wulff 1991). As seen earlier in figure 4.2, trophic exchanges have been estimated among thirty-six ecosystem compartments in Chesapeake Bay (Baird and Ulanowicz 1989). A similar description of the Baltic ecosystem has been made by Fredrik Wulff, who parsed that system into twenty-some components. The parties concerned soon discovered that both networks could be aggregated into virtually the same fifteen groupings, as shown in figures 7.5 and 7.6. One sees there that the topologies of exchanges, while similar, are not identical. It is the differ-

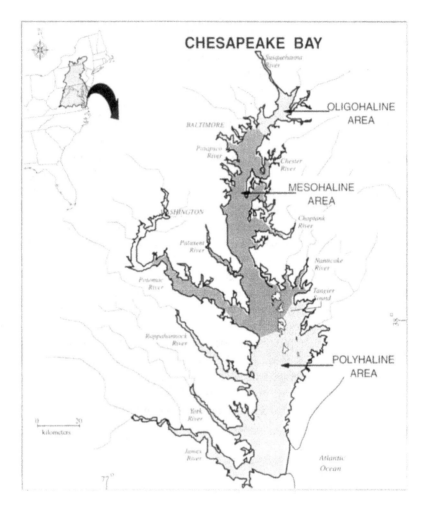

FIGURE 7.3.
Map of the mesohaline reach of Chesapeake Bay.

ences in the connections and their relative magnitudes that make the comparison interesting.

One obvious difference between the networks is that the Chesapeake ecosystem is far more active than the Baltic: its total system throughput (11,224 mg carbon/m²/d) is more than four times that of the Baltic (2,577 mg C/m²/d). Some of the higher productivity in Chesapeake Bay can be ascribed to warmer temperatures, but higher nutrient inputs to the Chesapeake are also likely to enhance its activity. Because TST scales

FIGURE 7.4.
Map of the Baltic Sea.

all the information indices, the ascendency and its related measures are uniformly greater in the Chesapeake than in the Baltic.

Of more relevance to the comparison are the unscaled values of the information indices themselves. The flow diversity was almost identical in both systems (4.50 bits in the Baltic vs. 4.47 in the Chesapeake), but the expression of development capacity differed in the two systems: almost 6% more development capacity appeared as ascendency in the Baltic than in the Chesapeake (38.4% vs. 32.6%). Most of the difference in relative overhead was due to the larger diversity of inputs to the Chesapeake, where allochthonous imports play a significant role.

These results suggest that Chesapeake Bay is more heavily perturbed (eutrophic) than is the Baltic. This conclusion was no surprise to investigators working on the Chesapeake. It raised some eyebrows, however, among Scandinavian ecologists, who were surprised to discover that a

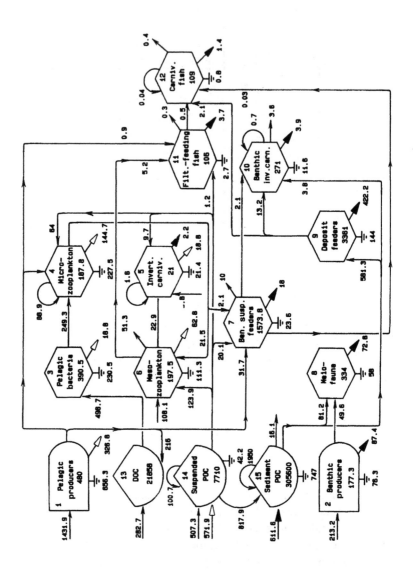

FIGURE 7.5.

Estimated flows of carbon among fifteen principal components of the mesohaline Chesapeake ecosystem. Flows are in mg/m²/d, and standing stocks
(inside boxes) are in mg/m². "Ground" symbols indicate respirational losses;
open arrows, returns to pelagic detritus; and filled arrows, returns to benthic
detritus. (*DOC* = dissolved organic carbon; *POC* = particulate organic carbon.)

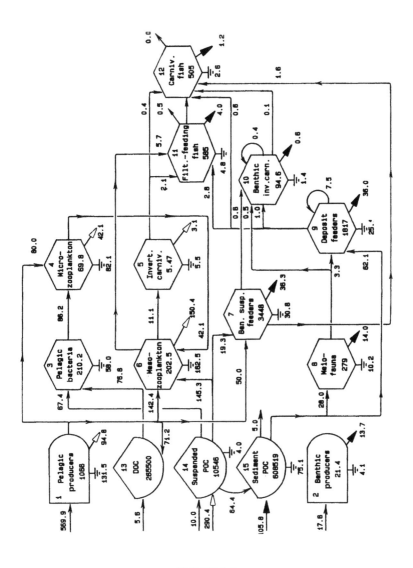

FIGURE 7.6.
Flows among the same fifteen compartments (see figure 7.5) as they occur in
the Baltic Sea network.

more haline ecosystem like the Chesapeake could be more disturbed than the Baltic. The implications of such quantitative comparisons for public policy should be obvious.

7.5   Corroborative Network Analyses

If making a judgment about the comparative trophic status of two ecosystems based on a few information indices seems like a risky business, such misgivings are entirely understandable. For this reason, the comparison between the Baltic and the Chesapeake ecosystems was buttressed by a broader analysis of the two networks. It is possible to search both networks, for example, to enumerate all cycles of carbon—that is, to identify all pathways in the trophic network whereby a quantum of carbon leaving a given compartment can be exchanged among the other compartments and return to the original node (Ulanowicz 1983). Peter Van Voris and his team (Van Voris et al. 1980) have analyzed data from several old field mesocosms that support the hypothesis that systems with greater resistance to and resilience from perturbation are also more complex, in the sense that they contain more and longer loops of connections that cycle at lower frequencies. Conversely, we would expect that perturbed ecosystems will possess fewer such cycles, due to sundry disruptions, and that, on the average, each cycle will involve fewer transfers. And indeed, cycles in the Chesapeake were deficient both in number and in average length.

Further support for the comparison was found in the trophic pyramids contained in the two ecosystems. The notion of a trophic pyramid follows from the application of the second law of thermodynamics to ecology. That stricture requires that each trophic transfer result in the loss of some finite amount of material and energy; hence, less medium will be available at higher trophic levels. This pyramidal effect would be obvious if ecosystems were simple chains of transfers, but most real ecosystems involve webs rather than chains of transfers—as a glance at any of the network figures in this volume will reveal. W. Michael Kemp and I (Ulanowicz and Kemp 1979), however, have described a mathematical mapping that apportions any particular trophic exchange according to how much of it is flowing after two, three, four, or more trophic transfers; this algorithm converts a complicated web of trophic exchanges into an equivalent trophic chain, or pyramid (Ulanowicz 1995c).

The trophic pyramids created by this process from the Baltic and Chesapeake networks are shown in figure 7.7. Because each trophic pathway is a concatenation, perturbations occurring at any point will disrupt all its flows to higher levels. We would expect, therefore, that the trophic pyramids of perturbed systems will be shorter and that less medium will reach the upper levels. This is precisely what the comparison shows: the pyramid derived from the Chesapeake network has one less step in it, and the relative amount reaching the last levels is far smaller than that flowing into the same level of the Baltic pyramid.

Metaphorically, the network of trophic exchanges in an ecosystem resembles the skeleton of a vertebrate organism. The comparison of the information indices between networks, then, when complemented by the analyses of cycles and inherent trophic pyramids, may be thought of as an exercise in the "comparative anatomy" of ecosystems (Wulff and Ulanowicz 1989). Together these methods comprise a significant quantitative approach to assessing the relative health of ecosystems.

7.6   Deductive Predictions from the Ascendency Hypothesis

In chapter 5 I mentioned that Liebig's "law of the minimum" can be deduced from the principle of increasing ascendency. Calculating the sensitivities of the ascendency to changes in the amounts of each element in the biomass of a taxon pinpoints which element most limits the growth of that taxon. In the Chesapeake ecosystem most of the lower trophic components were constrained primarily by nitrogen, whereas bacteria and vertebrate fishes were limited more by phosphorus. We now ask a follow-up question: "Of the several flows of a limiting nutrient into a compartment, to which one is the system as a whole most sensitive?"

Conventional wisdom suggests that the largest inflow of the limiting element should be the transfer that most affects system activity and organization. However, when the sensitivity of the overall ascendency to the change in any particular flow is calculated, the result is simple, but unexpected: it turns out that the controlling inflow of limiting nutrient is that which depletes its donor pool at the fastest relative rate (Ulanowicz and Baird forthcoming). This flow is not always the largest inflow of limiting nutrient. If the stocks of an element from one donor pool are being depleted relatively faster than those of another, then on a per-unit-biomass basis the resource is more readily "available" from the former. Control, therefore, is exercised by this donor, even though flows from other sources may be larger in absolute magnitude.

An example of such limitation by a secondary input is provided by the mesozooplankton dynamics in the Chesapeake ecosystem (Baird,

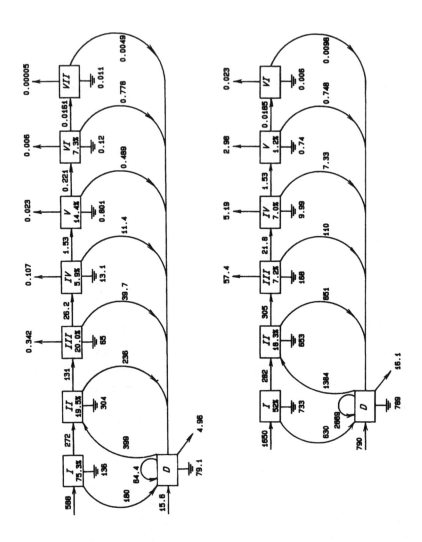

FIGURE 7.7.

The results of aggregating the flow networks in figures 7.5 (*bottom*) and 7.6 (*top*) into distinct trophic levels according to the algorithm of Ulanowicz and Kemp (1979). The numbered boxes represent abstract integer trophic levels; the percentage in each box quantifies the trophic efficiency at that level. All nonliving material has been lumped into the single compartment labeled *D*.

Ulanowicz, and Boynton 1995). Comparing the retention times of carbon, nitrogen, and phosphorus by the mesozooplankton, it is clear that nitrogen is retained the longest; hence, mesozooplankton dynamics are limited more by the availability of nitrogen than by either carbon or phosphorus. Now, mesozooplankton obtain their nitrogen from three sources: (1) phytoplankton, (2) the detritus-bacteria assemblage, and (3) microzooplankton (figure 7.8). (Mesozooplankton are usually reckoned as those planktonic fauna longer than 20–30 microns, and microzooplankton are the smaller animals in the plankton.) The largest input of nitrogen into the mesozooplankton flows from the phytoplankton at a rate of ca. 5,900 mg nitrogen/m^2/y. The donor pool of phytoplankton nitrogen averages 455 mg N/m^2 over the year. The smallest inflow of nitrogen to mesozooplankton comes from the microzooplankton at the rate of 1,720 mg N/m^2/y; however, this flow issues from a nitrogen pool that averages only 33.9 mg N/m^2 over the four seasons. Thus, the store of nitrogen in phytoplankton is being depleted by the mesozooplankton at a rate of about 3.5% per diem, whereas the microzooplankton yields its stock of nitrogen to the same predator at the much faster rate of 14% per diem. (The rate of depletion of detritus-bacteria nitrogen by mesozooplankton is lower than either of these values.)

Conventional methods of identifying nutrient limitation deal only with the aggregate amounts of various nutrients that are presented to the predator taxon. Such analysis would correctly identify nitrogen as the element most limiting to mesozooplankton growth. Liebig's law, however, provides no clue as to which source of that nitrogen is limiting. In the absence of any guidance to the contrary, the natural inclination is to rank the importance of various nitrogen sources according to the magnitudes drawn from each pool, and in most cases, this assumption accidentally identifies the controlling source. However, in this instance it would cite phytoplankton as the most crucial origin of nitrogen for mesozooplankton. Theoretical reasoning and hindsight point instead to microzooplankton as the key donor.

By identifying both the element most limiting to each species and its controlling source, it is possible to piece together a diagram of nutrient controls within an ecosystem. Figure 7.9, for example, sketches out the pathways of nutrient controls within the Chesapeake mesohaline ecosystem. Note that not all taxa are sensitive to the same elements. Furthermore, the most sensitive element may change (several times) along any particular concatenation of controls.

It is encouraging to see the principle of increasing ascendency yield a well-known "law" as one of its corollaries. It is more exciting still that

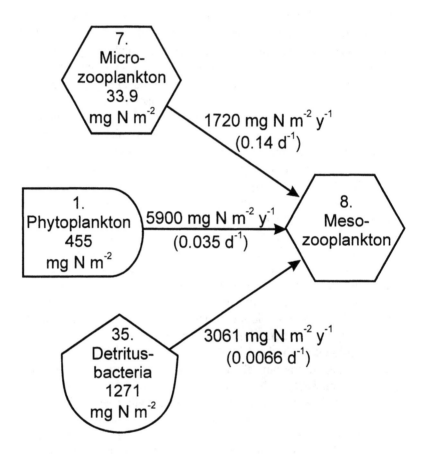

FIGURE 7.8.
The three sources of nitrogen that sustain the mesozooplankton in the
Chesapeake ecosystem (figure 4.2). The number in parentheses associated
with each input is the rate (per diem) at which that flow is depleting the stock
of that particular source (indicated inside the donor box).

the theory also provides a method for identifying controls in situations
for which no guidance currently exists. The relationship between increasing ascendency and Liebig's law demonstrates a necessary connection
between the new theory and accepted methods, and the rule for identifying controlling sources stands as a theoretical "prediction" against which
the ascendency hypothesis can be tested. That the predicted rule, in hindsight, makes good heuristic sense is most encouraging.

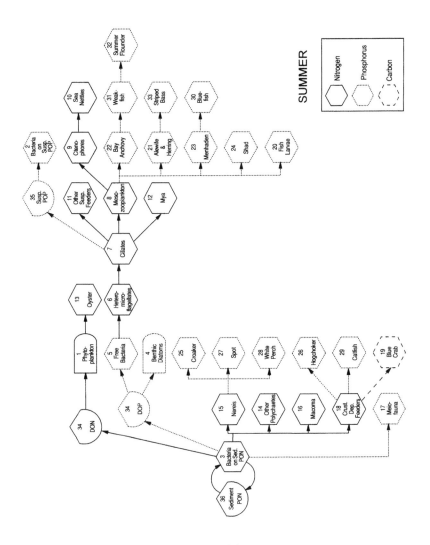

FIGURE 7.9.

A schematic diagram of flows (of carbon, nitrogen, or phosphorus) that control
each of the thirty-six compartments of the mesohaline Chesapeake ecosystem
(figure 4.2). The element that limits each component is indicated by either a
dashed (carbon), a dotted (phosphorus), or a solid (nitrogen) line; e.g., the
nitrogen in the ciliates (#7) controls the mesozooplankton (#8), whereas the
phosphorus in the latter controls the bay anchovy (#22). (*DON* = dissolved
organic nitrogen; *DOP* = dissolved organic phosphorus; *PON* = particulate
organic nitrogen; *POP* = particulate organic phosphorus.)

7.7   Noncognitive Values

With our calculations identifying the limiting elements and their sources within an ecosystem we are touching upon the issue of "value" in natural systems. Of course, "value" has many cognitive and normative associations, none of which are applicable to systems having no members capable of rational decisions. But, as many philosophers and sociologists are keen to emphasize, value, even in human situations, is not predicated entirely upon rationality. We are prone to think that values are assigned "from above" by the semiautonomous domain of the conscious. Koichiro Matsuno (1993) and Otto Rossler (1987), however, remind us that measurement (valuation) and interaction might proceed from within. The telos of an organism, as we inferred from our arguments on autocatalysis, is in the direction of acquiring the resources necessary for its own participation in the ecosystem. Those necessary resources that are scarcer in the environment will be sequestered by the participants for a longer duration. In this sense we speak of taxonomic compartments as "valuing" their limiting elements more than other resources. Furthermore, it is assumed that any ordination in value would reflect the relative magnitudes of the retention times of the resources at issue. (Cf. Simon's remarks reported in section 5.3.)

We have seen, however, that retention times are equivalent to the derivatives of the overall ascendency with respect to changes in biomass. In other words, the values that the system as a whole bestows upon various storages and flows are embedded in the ascendency. Furthermore, the embedding is not all that deep. We noted that ascendency has the form of a thermodynamic work function, which means it can be expressed as the sum of products of existing flows times their conjugate "forces" (which we generalized more aptly as "propensities"). Later we saw that whenever flows are expressed as energy, their logarithmic cofactors represent the increase in the quality of the energy as it enters a higher trophic compartment. That is, the product of the energy exchange times its increase in quality is equivalent to the rate at which exergy is being stored in the receiving component. The ascendency, being the aggregation of all such products, therefore represents the overall rate at which the system is doing work (i.e., storing exergy). We then sketched out a numerical scheme to estimate the magnitudes of a "potential" function pertaining to each node (see section 6.3).

In a formal sense the magnitude of this potential at each node should correspond to the "value" of the stocks stored there. This value is not the result of any cognitive decisions, as in traditional economics. Rather, it estimates the value of that particular resource *in the context of* the functioning of the entire ecosystem (i.e., *sensu* Matsuno). These remarks might interest those connected with the burgeoning discipline of "ecological economics" (Costanza, Daly, and Bartholomew 1991; Ayres 1994). A primary goal of this new endeavor is to express the economic values of those biotic resources that normally lie outside traditional markets. What intrinsic value, for example, can one place on the insects in a forest floor, or the zooplankton in an estuary? Normally these populations do not enter into human economic transactions, even if they do participate in the natural "marketplace" of the ecosystem. For example, as a resource necessary to a population becomes scarcer, the taxon usually responds by sequestering that element for a longer time; through this change in behavior the organisms are adjusting the degree by which they "value" that resource in the face of altered market conditions (availability in the environment).

Ascendency provides a way to estimate the relative values of biotic stocks as members of a functioning ecosystem. If, in addition, two or more biotic components should also have values within the human marketplace (e.g., trees for lumber and fuel, fish for food and sport, mammals for fur, etc.), then these taxa become the keys or "Rosetta Stone" for converting relative values within the ecosystem into monetary contributions to the human economy. Such translations have been attempted before (H. T. Odum 1988; Ulanowicz 1991); in those efforts, however, the relative values within the ecosystem were based upon the amounts of material or energy per se that were embodied in usable exports from natural system. Ascendency, being a work function, serves as a closer analog to the "capital generation function" in economics. Accordingly, there is more reason to believe that the relative values of resources will be better assessed by the new technique.

As with earlier attempts to translate ecological value into monetary terms, one must take great care to view the estimates as conservative. For example, the market value of trees, when used as a conversion factor, reflects only the value of the wood to humans as lumber and fuel; it does not include trees' contributions as producers of oxygen or moderators of temperature, etc. Virtually any attempt to incorporate biotic "externalities" into the market seems destined to ignore important biotic contributions to human welfare.

7.8 Applying Ascendency to Traditional Models

The fact that information indices derive from probability distributions and not from mechanical considerations renders these measures more broadly applicable, as we have seen in the case of Langton's cellular automata. Their generality, however, is no reason to ignore the relevance of the indices to phenomena that are particularly mechanical in nature. For example, in constructing a mechanical model of an ecological system, it may prove useful to take a "snapshot" of the model flows and biomasses from time to time. Calculating the components of the ascendency and related variables at intervals during a model run can provide the modeler with insights into how the mechanical analog is behaving.

John Field and his team (Field, Moloney, and Attwood 1989), for example, created a seven-compartment model of the microplankton community in the Benguela upwelling current. They simulated the development of the system over a forty-day period and took nineteen "snapshots" of the trophic flow network as it changed over the interval. Field noted that the magnitudes of ascendency and related variables were driven strongly by the system growth (increase in TST). Nevertheless, the mutual information of the model network (the ascendency divided by TST) evolved in roughly sigmoidal (smoothed-steplike) fashion. Mutual information declined slightly beyond day 20, indicating that the system perhaps had begun to senesce as it ran out of resources. Ecosystems in upwelling regions of the oceans play out under particularly strong physical and chemical constraints. The temporal behavior of Field's calculated information variables reveals that he captured the controlling forces with a fidelity sufficient to portray the full course of ecosystem development.

There is not always such obvious accord between the output of mechanical models and the prescriptions of ascendency theory. Whenever the two disagree, a common inclination is to regard the discrepancy as evidence *against* the ascendency hypothesis. (Vestiges of our Newtonian heritage are slow to fade.) When this temptation arises, one should remember that information theory was invoked specifically to describe and quantify phenomena that are *not mechanical* in nature. Hence, all other things being equal, there is at least as much reason to question the adequacy of the mechanical model as to doubt the relevance of the ascendency hypothesis.

This phenomenological priority of the network approach suggests the use of independently derived trophic flow networks as benchmarks against which to compare the outputs from a mechanical model of the same ecosystem. For example, the U.S. Geological Service currently is supporting the development of a highly complex "multi-model" of various South Florida wetlands (Fleming et al. 1995). The planned Across Trophic Levels System Simulation (ATLSS) will be generated by a composite of modules, each of which mimics the behavior of a particular ecosystem component at an appropriate level of resolution. Thus, the larger fauna, such as alligators, deer, and wading birds, are being modeled down to the resolution of each individual organism (DeAngelis and Gross 1992; Wolff 1994). Forage resources, such as fish, crawfish, and amphibians, are to be simulated in terms of age- or size-structured populations using Leslie (1945) matrices. Finally, primary resources, such as macrophytes, periphyton, and detritus, will be represented in the context of coupled differential equations as pools of homogeneous biomass. After the numerous modules have been placed within an algorithmic "shell" and the ensemble has been run, the model can be queried at convenient intervals to produce a time series of trophic exchange networks. These outputs can be compared with the independently derived networks as a form of model verification. As a bonus, should the interactions among the ATLSS modules produce some strange and possibly unrealistic behaviors (as should be expected initially), the suite of network analysis routines described in section 7.5 can be applied to the benchmark networks to help illumine where the model is going wrong.

7.9   Neural Networks

In chapter 4 we noted how the human autonomous nervous system continues to mature even after the cessation of any increase in its development capacity. Recent years have seen considerable efforts to create mechanical analogs of neural processes, models commonly referred to as "neural networks." One of the more physically detailed analogs is that by John Hopfield (Hopfield and Tank 1986). Hopfield presented a set of differential equations that describes the electric currents flowing among an interconnected ensemble of model neurons. Each neuron is characterized by its interaction strengths with other neurons, its capacitance, leakage resistance, and firing rate. The last characteristic is portrayed in terms of a steep sigmoidal (virtually steplike) function of the applied voltage, so that each neuron at any instant is effectively in an "off" (low voltage) or "on" configuration.

It can be demonstrated that whenever the strengths of the interaction coefficients between neurons are symmetrical with respect to each interacting pair, the Hopfield equations always converge to one of a finite set of steady-state solutions. Each solution consists of a combination of neurons in the off and on positions. Furthermore, each steady-state solution (stationary point) possesses its own finite "basin of attraction." The result is that for any initial condition that is within the basin, the system eventually will wind up at the corresponding stationary point.

This behavior is extremely useful for implementing pattern recognition by mechanical devices. Each stationary point can represent a discrete canonical pattern (as described by a string of ones and zeros). If a starting pattern that is not identical to one of the canonical forms (stationary points) is fed into the network, the system will converge to the center of whatever basin contains the starting point; that stationary point is then read as output. In effect, any "blurred" input pattern will be "recognized" as its nearest "sharp" canonical form.

Because the Hopfield equations actually mimic transient voltages and currents in the electric networks, it becomes possible to portray the system at any instant as a network of electric currents. Corresponding to each such "snapshot" one may calculate a system ascendency (and related information variables). Alternatively, one could calculate the temporal ascendency of the entire time series of snapshots (Pahl-Wostl 1992). Kenneth Bosworth and I (Bosworth and Ulanowicz 1991)[1] have tried to ascertain whether the ascendency of a neural network reflects its ability to interpret input patterns correctly. In particular, we performed progressive "brain surgery" on the networks by successively excising interneuronal connections. The neural networks were surprisingly robust in their ability to recognize patterns, even after the removal of major interconnections that we had anticipated would render the system dysfunctional. When collapse finally did come, it was relatively sudden. For as long as the system retained its functional capability, any drop in its ascendency due to excision was only incremental. Near system collapse, however, the temporal ascendency went into perceptible decline.

We saw these preliminary results as encouraging and had planned further experiments to test the relationship between ascendency and "learning" by networks. One "teaches" a network to respond in a specific way to given inputs by providing feedback to the interaction coefficients. That is, whenever the system responds with the desired outcome, those interaction coefficients that contributed most to the solution trajectory are amplified by a small amount, and whenever an outcome is other than that desired, the contributing interaction strengths are attenuated slightly. Gradually, the system comes to respond in the desired way most

---

1. This citation is to the abstract of a presentation at a meeting of the American Mathematical Society. Unfortunately, soon after the meeting all computer files pertaining to the analysis were destroyed in a massive computer crash. For a multitude of personal reasons, the work was never repeated.

of the time. We intended to follow the ways in which the temporal ascendencies of the networks changed as they were trained. Our expectation was that as a neural network learns a task, the temporal ascendency of its transient flow structure rises. Furthermore, we anticipated that the plateau in magnitude it achieves after training has been completed will vary monotonically as the complexity of the task that has been learned. In other words, we envision the network ascendency as an index of performance for the neural net.

7.10   Computational System Performance

System ascendency can serve as an index of performance for other systems as well. For example, today large complex tasks are often executed in distributed fashion among a network of interconnected computational machines. Flows between pairs of machines in the net can be recorded in some convenient unit of communication, such as megabytes per unit of time. The temporal overhead associated with the time series of such communication networks should provide a measure of how inefficiently the task was performed. Usually, distributed computation is accomplished under the aegis of some central controller. But new configurations employing distributed control are being studied (Hogg and Huberman 1991), and temporal overheads corresponding to different strategies for network control could be a highly effective index with which to rank relative performance.

There is little doubt that the opportunities to apply information measures like ascendency will continue to multiply. As the scope and utility of information theory continue to grow, so should the number of investigators and philosophers who accept the major shift in causal perspective that Popper advocated. The ramifications behind his new paradigm for evolution run wider and deeper than its particular applications. It is only fitting, therefore, that I devote the last chapter to considering the broader implications of how the combined visions of Karl Popper and Robert Rosen could lead us into the next century.

8

THE ASCENDENT WORLDVIEW

8.1 Brittle Attitudes

Having considered how ecosystem science appears to demand non-Newtonian descriptions, I now step back to contemplate how this new perspective may affect our vision of the wider world. What follows is a litany of issues that could be affected by the Popperian viewpoint in general, or by the concept of ascendency in particular. Most assuredly many, if not all, of these insights have already been discussed at length by professionals in the appropriate fields, and I beg the indulgence of anyone whose particular contributions I have yet to discover. The goal here is not to be exhaustive, but simply to indicate a few directions in which scientific thought and discourse might legitimately proceed once it has been freed from the strict confines of the Newtonian paradigm.

Let me begin by reviewing the only picture available to us if we make the assumption that all causes are material and mechanical in origin. That image is both rigid and universal. For rigidity is the crux of mechanical behavior: If A, then B—no exceptions! Universality follows from it. If some outcome other than B should happen to transpire, then surely it was the result of intervention by some other mechanical object or event. The world immediate to our senses becomes one made up entirely of events that are tightly coupled to each other. If matters should appear otherwise to us, we are told it is due only to our ignorance of details and to our imprecision in quantifying objects and changes. The assumption is that if complete knowledge and precision were available to us, we should be able to predict the effects of any given event far into the future and at places quite remote.

A world of events so linked up is causally "closed." Nothing truly novel can occur in one's immediate surroundings. In pursuit of the origins of events, one is led on a causal regression either down the hierarchy toward elementary particles, or out into space. Hence the popular image of "fundamental" research as consisting solely of the activities of the particle physicist, who is searching for the "essence of reality" among the quarks, or of the cosmologist, who reads the "secrets of the universe" in light that arrives from the farthest reaches of space. To suggest that an ecologist or an economist deals at times with the fundamentals of natural existence is to inflict mild embarrassment upon a polite audience, or to risk open derision from the ostensibly confident and secure.

The allure of finding an unambiguous single cause behind each and every event is great, of course. It reflects our belief that nature, at its core, is simple and knowable by us. In expanding the frontiers of observation at both extremes of scale, we hope eventually to achieve a vision that will merge all four types of physical forces into a single unity. One principle then would guide all science, and every event would be traced back to its origin in the one unifying source.

It should be acknowledged that the fruits of the Newtonian approach have been enormous. Who can but marvel at the power that Newtonian laws give us to insert satellites into orbits around distant planets? Who can question that the reductionistic imperative is responsible for the discovery of bacterial and viral agents that cause so much human suffering? Nor should anyone deny that the identification of the material locus for the genetic transmission of information ranks among the major achievements of this century. Francis Bacon's motto reigns supreme: "Scientia potentia est"—knowledge is power. We may bask in the possibility that there are no limits to the control that science affords us to exercise on the world and ourselves. Like Dr. Faustus, we have become as gods among the heavens (Faber and Manstetten 1991). We need only acknowledge in return that material and machinery are the ultimate and only elements that populate our universe.

Given such achievements and expectations, why should anyone wish to question the prevailing view? Certainly, the few meager results offered in the last chapter pale in comparison to what the Newtonian approach has wrought. In the end, however, one single observation makes it impossible *not* to question the sufficiency of the Newtonian description: *Life itself cannot exist in a wholly deterministic world!* Some degree of causal openness is essential if living systems are to respond to new and novel material and the energetic signals that continually threaten them. In closing the "treaty" we now call the "modern evolutionary synthesis," biologists *reluctantly* had to make room for sto-chasticity—but they hardly cracked the door (Depew and Weber 1988). Strange and uncontrollable things might happen during the moment of meiosis, but once that unpleasantness is past, we again turn our attention toward survival in a Newtonian world that imposes strict conditions on survival and propagation, and rewards only those that optimize their reproductive and adaptive trajectories.

In the world that ecology now invites us to consider, no ultimate refuge from the accidental is possible: chance configurations of processes can arise and, once in existence, can actively change the surrounding world. But ecology does not allow chance to operate unconstrained. A chance event at any particular scale cannot ramify indefinitely up and down the hierarchy. Particle physicists like to remind us that changes in the fine structure of matter would make for a wildly different world. Perhaps this is true of a few key relationships (e.g., the molecular structure of water), but there likely are innumerably more behaviors at small scales that could be other than they are without having any appreciable impact on organization at higher scales.

Popular attitudes toward science often find expression in science fiction. One of the favorite themes of this genre is time-travel, wherein characters who matured in one era suddenly are transported back to a previous age. In plots built around this device, like those by H. G. Wells, tension often is generated by the explicit necessity that the time-travelers not alter anything in the past, lest history in the interim run amok. The extent to which readers are inclined to accept this condition reflects the wide-spread belief in strict determinacy across all scales and all levels. One little perturbation and the entire house of cards collapses. In a causally closed world there are no true alternatives: The kingdom was lost in a decisive battle. The battle was lost because the king did not arrive in time to lead his legions. The king was delayed because his horse had not been shod. The blacksmith was frustrated because he had run out of nails. Ergo, for want of a nail, the kingdom was lost! Most of us smile at the way this familiar tale is overdrawn—but as regards the nature of dynamics, we usually acquiesce in long, unbroken chains of mechanical forces just like this.

Forces, however, are agencies that act only in isolation. As Popper suggests, they are at work only in the vacuum of space and the sterility of the laboratory. Elsewhere, for any event to have an effect somewhere removed, it must act through a welter of other interfering, unpredictable phenomena. Hence, in everyday circumstances, it usually is better to speak of propensities than of forces. In that sense it is folly to talk about genes as *determining* social structure and behavior. This is not to say that the current genetic makeup of a population does not influence how it behaves—it is just that, over time, genes do not set their own agendas. They dance to the tune played by larger, formal agents.

8.2   Adaptable Perspectives

In sum, the world is open, not deterministic or rigidly coupled. It does not become utter chaos, however. A fundamental asymmetry arises whenever many different elements interact with each other. Most such interactions are arbitrary and lead nowhere—they come and go. Inevitably, however, some configurations of interactions become mutualistic, self-reinforcing, or, as Robert Rosen (1991) puts it, self-entailing. They grow and persist longer relative to their more arbitrary surroundings. In time, their kinetics impart an overall structure to the system, acting as the cohesion that builds and maintains order.

The appearance of order is not so exceptional an event as the second law, in the way it is usually stated, would have us believe. For example, one rendition of the mandate is that it is impossible in any irreversible process to convert a given amount of energy entirely into work without rendering some of it useless. The connotation is that order is contingent, and dissipation, inevitable. Nothing prevents us, however, from casting the obverse and noncontradictory statement: "In any real process, it is impossible to dissipate a set amount of energy in finite time without creating *any* structures in the process." Often the structures (work) created are quite ephemeral, but they are as inevitable as dissipation.

Not only is the appearance of structure ubiquitous, but, once having arisen, it can function as a cause in its own right (the formal causality of Aristotle or the "downward causality" of Donald Campbell [1974]). Configurations of processes, as we have seen, are able to change not only the material and mechanical makeup of their surroundings, but, over time, their own composition as well. None of this implies that formal agency exists apart from its material constituents. It implies only that the behavior of a complex system is not fully dictated by its components. That is, the actions of formal agency are, to a degree, *autonomous* of those of its constituents.

The asymmetry that gives rise to structure also specifies a preferred direction. When asymmetry is compounded with the autonomy inherent in formal agency, the result is a self-direction, or telos, proper to the system (the final cause of Aristotle). With the introduction of telos into ecological narrative, departure from the Newtonian worldview becomes more radical. Instead of a world that is one uniform material continuum connected in rigid fashion by mechanical links, the ecological arena is populated by discrete entities, each with its own direction. While this might seem at first like a prescription for rank disorder, the direction of each agent can mesh with those around it, just as do each agent's internal constituents separately. The results are metastructures in the larger realm. At each level of the hierarchy, agencies inevitably arise to structure events.

This new perspective reveals a more realistic balance to the natural world. The Newtonian conception cast the world as a perfect dynamical construct, without flaw or imperfection. Somehow endowed with motion, the world was compelled to play out its script in clockwork fashion. This flawless construct was then disrupted by the criticism of thermodynamics, which in its turn condemned the world to the disorder of heat death. Common sense tells us, however, that both order and disorder are necessary attributes of reality. The thermodynamicists are right—the world is not changeless, as the symmetrical laws of Newton would imply. But the appearance of disorder is not the only real change that characterizes a dissipative world: new order is inevitable as well, and forms appear that are only partly explicable by reference to their composite events. The world as we perceive it is the outcome, then, not of a static balance, but of a balanced conflict—the opposition of propensities that build order arrayed against the inevitable tendency for structures to fall apart. Just as with a true dialectic, both sides in the contest require the existence of the other. Dynamical order cannot be maintained without the buffer afforded by disorder; conversely, forms cannot disintegrate without creating at least transient structures that serve as potential seed for new order.

Many will object that the new perspective sounds too much like the transcendent narratives that antedate the scientific revolution. Over the past three centuries we have developed marvelous tools to extend our senses into previously unknown realms. To draw coherent pictures of the world at remote scales, it has been necessary to adopt concepts and attitudes that often were at odds with common sense (i.e., our image of the world conveyed by the immediate senses). That this exercise has yielded some insights into the proximate world is not to be questioned. By and large, however, it has not sufficed to paint a coherent picture of events at hand. Natural ecosystems constitute a significant part of the immediate world. It should not be surprising, then, that to make sense of ecology we need to reexamine some attitudes that were discarded in our efforts to understand more distant realms. There is no reason, furthermore, to regard knowledge that arises from immediate experience as necessarily inferior to that obtained via mechanical aids. These considerations notwithstanding, few scientists are likely to abandon their cherished beliefs about the fundamentals of nature on the basis of propositions advanced by those interested in birds, bunnies, and bugs. After all, everyone knows what material progress can be achieved by viewing the world as a machine. In addition, the question of whether chance is ontological or epistemological remains, ultimately, a matter of definition. What harm, therefore, can come of continuing to believe in a world as seen through the eyes of Laplace's demon?

What harm, indeed? The danger in refusing to adopt a hierarchical perspective on nature is that, by clinging to the clockwork analogy, we will preclude from our narratives all but a limited fraction of the phenomena that constitute our daily world. If Popper is correct, and forces are but a small subset of more ubiquitous entities called propensities, then we would forever remain blind to the workings of the latter if we were to remain faithful to the Newtonian-Laplacian cosmology. That is, we would intentionally forgo the possibility of studying the effects of formal agency and telos in any quantitative and systematic manner.

8.3   Medical Insights?

One need look no further than the field of medicine, where the analogy of the body as machine has stimulated most progress to date, to find examples of how such ignorance might impair human welfare. There are parts of the body where the machine analogy works quite well—the heart and circulatory system, for example (see, however, Cesarman 1996). The heart as a pump can be repaired, and can even be replaced for a while by a mechanical substitute. Vessels can be cut, spliced, reamed, and otherwise treated as one would repair plumbing conduits. Several hours spent in the waiting room of a hospital's cardiac unit should be enough to convince anyone of the efficacy of this approach: most of the news brought to patients' relatives is upbeat and encouraging.

Optimism does not run as high, however, in oncology units. Cancer does not lend itself as readily to mechanical analogy. If a growth is detected early enough in a nonvital area, surgical excision often provides a remedy. But indications are that cancer is a system-level affliction, and the efforts at a cure that seem most promising involve some manipulation of the immune system, either with drugs or via the stimulation or infusion of specific classes of antibodies.

Perhaps because of the successes of medicine in pinpointing the etiologies of many diseases as small-scale invasive pathogens, research on carcinogenesis seems to emphasize the search for eliciting microagents. Some investigators look for slow-acting viruses that eventually might trigger uncontrolled cell division. Others search for "oncogenes," or genetic defects that predispose an individual to cancer. Still others study chemical or biological toxins that render the body vulnerable to metastasis. None of these agents, however, emerges as a clear efficient cause in the sense that whenever it occurs, it always engenders the disease. Each agent has a *propensity* to induce the disease, but none *forces* its onset. Often, individuals exposed to one or several such factors over a long time fail to develop the illness.

The preoccupation with microscopic agents is especially puzzling when one considers the preconditions for the onset of cancer. J. L. R. Chandler (pers. comm.) cites four events that must transpire before the organism is threatened:

1. A mutant cell must appear.
2. The growth must escape from the local matrix of like cells in the same tissue.
3. It must evade the surveillance of the immune system and the hormonal control of central homeostatic processes.
4. It must successfully invade a foreign tissue and compete for resources in the new environment.

It is obvious that only the first step involves a microscopic or point event; the remainder encompass system-level or subsystem activities. Yet most cancer research focuses on microscopic etiologies. Not that we should ignore the roles of microscopic agents, for they most certainly play a part—but to come fully to grips with the nature of the disease, we probably need to emphasize more the system-level nature of the pathology.

The immune *system* is far more difficult to understand than the workings of a virus or a gene. It varies among individuals for reasons having to do with both heredity and life history. It is not conveniently represented as a mechanical entity. That the immune system exhibits a degree of autonomy from control by the host organism is sometimes made painfully evident in the allergic and autoimmune reactions that plague so many.

It might be fruitful to consider whether cancer is a pathology of the immune system that, once established, pursues a course relatively independent of the eliciting insults. This possibility has been made clear by recent research into another nemesis of our times, acquired immune deficiency syndrome, or AIDS (Garrett 1993). Until recently, most work on AIDS has focused on the human immunodeficiency virus, or HIV, and its ability to harm the CD4 cells, a component of the human immune system.

Certainly, these are important things to know. But a consensus is now emerging among investigators that, from the earliest moments of infection, HIV tricks the immune system, setting it on a path to self-destruction that may continue *even if the virus is subsequently eliminated.* The major damage, some researchers say, is not the direct result of the virus, but of an immune system that has gone haywire, attacking itself and other parts of the body. In other words, it may matter a great deal whether one regards AIDS as the result of its admitted efficient cause (the HIV virus) or of its formal cause (a pathology of the immune system). In the former case, treatment would center on blocking reproduction of the HIV virus and preventing opportunistic infections, such as pneumonia. Under the latter perspective, more effort would be devoted to treating the immune system directly—using such tactics as jolting it early in the infection to set it back on its proper course, or hyperstimulating in it a sharp allergy-like response with such things as poison ivy, foreign tissue, or other irritants to override the self-destructive pathway initiated by HIV.

No one is pretending that research into ecosystem behavior is going to inspire a cure for cancer or AIDS. But we have found it plausible and convenient to identify organisms as superecosystems. If ecosystems cannot be understood by treating them as machines, then might not organisms be other than fully mechanical? Observing how ecosystems function has led to an entirely new perspective on nature. It could be that the ecological viewpoint, when applied to organisms, would suggest new approaches to ontogeny and pathology that otherwise would never have come to mind.

8.4   The Funding of Science

Medical research is only one area of science wherein the ecological perspective might reveal new avenues of inquiry. Many investigators working in the life and social sciences suffer from "physics envy"—i.e., the ardent desire to see their disciplines become hard, predictive endeavors, like classical physics (Cohen 1971). Unfortunately, the envious pay scant attention to the fact that contemporary physics has come to terms with life in an indeterminate world. Even less acknowledged is the fact that, when physics in the last century faced its crisis over determinism vs. chance, quantitative tools that had been created in the "soft" social sciences (probability theory) were what came to its rescue (Depew and Weber 1994).[1]

---

1. I might point out further that the physical sciences are not as "hard" as social scientists and humanists usually think (Porter 1986).

Today science seems to be directed less by the philosophies of investigators than by the prevailing attitudes of those who fund the enterprise. Of necessity, perhaps, funding agencies are guided by a very conservative philosophy of science. Virtually every sponsor of American science requires that submitted proposals be cast in the positivist format of falsifiable hypotheses. Although this requirement does impose a certain rigor upon how ideas are presented, it tends to cut both ways—for the positivist formula goes hand-in-glove with the reductionist view of nature. It discourages the inductive and phenomenological efforts that now appear necessary if we are to gain deeper insights into the living world. Even in ecology, we are forced to pay lip service to the belief that the world at hand follows the same scenario as the events we see through microscopes and telescopes. We often wind up with mere rhetoric rather than a statement of what needs to be funded.

The lady mentioned in chapter 1 who was puzzled by what a theoretical ecologist *does* has ample company. To see what little visibility theoretical ecology has, one simply has to look at the disparity between the issues that we profess are important and the research that we actually support. There remains widespread concern that the world may be facing ecological Armageddon. One hopes that, in the shadow of looming catastrophe, little will be spared in the search to avoid coming to a bad end. During the Cold War, when we were confronted with the prospect of nuclear annihilation, the American government poured hundreds of millions of dollars into research on particle physics, a reasonable fraction of which was earmarked for theoretical physics. But our response to the threat to our ecosystem has been quite different. We are rightly willing to appropriate billions for pollution control and cleanup, and millions more for research into global climate change. But if we were truly serious about avoiding catastrophe, we would be passionately interested in learning more about how ecosystems actually function. We would mount a credible program in *theoretical ecology*. As it now stands, the richest nation in a world faced by ecological upheaval commits less than one million dollars each year toward theoretical ecology—which is absolutely *nothing* on the national scale of things. In light of the gravity of potential consequences, this neglect transcends the merely scandalous.

8.5   Free Will?

Ignoring something that might save us is hardly the only irony of contemporary life. The American nation was founded upon the freedom and rights of the individual. Yet few contributors to the science that sustains this society give much credence to the idea of individual free will. This is because most scientists interpret "free will" to mean conscious action that is not fully determined by material and mechanical antecedents; and because it is a particular form of telos, there is no room for "will," free or otherwise, in the Newtonian scheme of things. The ecological viewpoint, however, holds that telos and autonomy can arise naturally in living, developing systems.

It is interesting that most arguments bearing upon free will vs. determinism are couched in terms of individual decisions, presumably occurring over short time spans of, say, a tenth of a second or so. It is when actions are observed under such short intervals that the determinist's arguments appear strongest. The depiction is usually akin to that of neural nets (section 7.9), wherein electrons, in response to an input, course over a network of synapses toward a final, predetermined configuration. Given a short enough time span and abundant information one can, in principle, always trace backward in time through a causal chain of mechanical events to make the final outcome seem inevitable (that is, determined).

Now, if the neural network is very highly articulated—i.e., if it is sparsely and strongly connected (has a high ascendency and low redundancy)—then the interval over which one can accurately hindcast past events and predict future consequences may be long enough to sustain the determinist's conclusions. Living networks, however, are unlikely to have vanishingly small redundancies. Redundancy in the neural system must be low enough to allow for coherent thought and response, but the network cannot be so highly articulated that learning (adaptation) is precluded. Given any reasonable redundancy and a large number of neurons, the characteristic time over which prediction remains possible becomes vanishingly short. The determinist's scenario quickly evaporates.

Free will, it would seem, pertains to a longer time frame. It is associated with a deliberative process. The common perception of free will usually is not associated with split-second decisions, which most agree can be regarded as prewired reflexes; rather, the exercise of will appears to require minutes at the very least, and often is the culmination of a very long period of induction. Based on what I have already argued, even a network of almost rigid, nearly mechanical components can exhibit feedback at the larger system scale, so that its trajectory over the long term derives from a formal cause acting in a particular direction (telos). This higher-level activity exerts its influence upon the neuronal components and their connections in a *selective* way. Over time it alters and replaces them in accordance with its own direction.

Put in less-abstract terms, a neurophysiologist can shock, cut, drug, or otherwise perturb the brain of a subject, and elicit from it a more-or-less predetermined response. The manipulator is seduced into thinking that the brain is a machine (Franklin 1983). Leave the same brain in relative peace, however, subjected to less-invasive stimuli over a longer time, and the resulting behavior becomes impossible to predict. Similarly, an individual may be thrown into disarray by a stroke or some other organic dysfunction, but such subjugation of mental activity to physical causes should not detract from the significance of what the same person may have accomplished over a lifetime of acting as a free agent.

8.6   Economics and Politics: Individuals and Societies

Even should an individual be free to choose his or her way in the world, such autonomy is obviously constrained by the propensities of society as a whole. To investigate further the relationship between the individual and society, let us look to the scenario for the development of ecosystems sketched in section 4.8 to provide us with analogies for the economic, social, and political development of societies at large.

Parallels between the course of economic development and the development of ecosystems are perhaps most evident in the evolution of economies in recently settled regions of the globe, like that of North America. In the beginning, the categories of economic activity (occupations) were relatively few: there were mostly farmers and soldiers, with only sporadic artisans and officials. Growth was manifested largely as an increase in total throughput, fueled by an expanding population and areal extent of cultivated land. In Europe during preindustrial times, when usable energy was derived from recent sunlight, the agricultural capacity of the land eventually became saturated. Depending upon the fertility of the region, some diversification of trades ensued, but stiff competition for scarce resources resulted in rigid and stratified relationships among the occupations of feudal society. Before matters ever reached this point on the North American continent, however, new energy was drawn into the economy in the form of fossil fuels. Not only did this infusion drive up the scale of activity, but it also served to prolong the increase in the diversity of occupations. (Recall that a larger throughput can be divided in more ways, before some units become so small that they risk extinction.) There being relatively little competition for resources, egalitarianism among occupations remained high and provided a fertile field in which democratic traditions could flourish. As American industrial society matured, specialization increased apace, both among individual occupations and in corporate activities (monopolies). Too much ascendency (specialization) relative to overhead (egalitarian generalism) is not, however, a stable situation, so recent decades have witnessed a proliferation of controls to balance and maintain the fruits of production—i.e., occupations like the law, police work, bureaucracy, health work, union organization, and politics. (A healthy, or mature, system has both a high ascendency *and* a high overhead.)

Whether a society is characteristically egalitarian or authoritarian seems correlated with economic conditions, which in turn drive much of politics. In fact, the very dialogue of politics reflects the larger conflict in nature. Those inclined toward reinforcing class structure, improving efficiency through specialization, and conserving wealth and the status quo (in general, the political right) comprise the agencies that build structure and increase ascendency. Those who emphasize social equality, breadth of outlook, tolerance, and change (the left) lend long-term stability to society through contributing to overhead.

The qualifier "inclined" and the verb "emphasize" just used underscore the fact that in the social dialectic, as in nature, significant representations from the opposing communities are required if a society is to remain a potent, functioning entity. The necessity of the opposition is tacitly assumed in the respective ethics preached by the antagonists. Ethics imply limitations on behavior. On one hand, the push toward ever greater ascendency, if left unchecked, would result in a clockwork society, wherein individuals become obligate cogs, or slaves to their functions in production. Semiotically recognizing the danger in its own ends, the political right fervently espouses protecting the rights of individuals. On the other hand, those on the left implicitly acknowledge that the unbridled implementation of their ideals would cause society to cease production and collapse. The ethos of the left, therefore, highlights the obligations of the individual to the community. In a healthy society ethical checks keep the dialectic from going too far in either direction and maintain the community within its "window of vitality." (This counterpoint of ethos with praxis also helps explain the irony whereby extremists of either ilk sometimes find common ground with each other: each works to bring about what the other preaches.)

Correlating the political right with ascendency and the left with overhead prompts some tenuous generalizations as to how political winds might shift with changing economic climate. In times when resources are abundant, little premium accrues to efficiency and conservation, so one expects the body politic to move leftward. Conversely, as resources decline, greater reward comes from efficiency and conservation, prompting a corresponding drift to the right. The sea change from postwar America to the present tracks this shift.

It is commonly agreed that excursions too far to the right or left can destabilize a society by moving it too near the boundaries of its "window of vitality" (section 6.6). It remains an open question how long any given polity can continue to persist within the window. Macroeconomic conditions are known to exhibit long-term oscillations. Presumably, an adaptable and sufficiently democratic institution should be able to adjust to such changing circumstances by shifting its political temperament accordingly. Some sociologists, however, remain less optimistic and posit an inexorable drift toward collapse.

Joseph Tainter, for example, elaborates the full scenario of social development—beginning with Michael Harner's observation (1970) that sudden access to new resources usually simplifies social structure, continuing through the gradual development of constraints to regulate and distribute those additional resources, and culminating with the collapse of the polity due to a "declining marginal return on continued investment in complexity" (Tainter 1988:120). It is not entirely clear how this marginal return connects with the information inherent in the flow structure, but it likely correlates with the fraction of a system's development capacity that is manifested as overhead. Figure 4.9 reveals that this fraction, represented by the vertical distance between the two curves, shrinks steadily once the system enters its mature stages.

According to Elam Service (1960), this drift represents the "law of evolutionary potential"—or, more accurately, the "law of *declining* evolutionary potential." For Service's law closely resembles Holling's progression toward "brittle" system structure: the gradual conversion of development capacity into ascendency, until the point is reached where insufficient potential remains to allow the system to adapt to new exigencies (Holling 1986).

We wind up recapitulating the arguments for and against Holling's emphasis (1986) on "creative destructions." Can a particular system remain in the upper right-hand corner of Holling's figure eight for an extended time with only mild excursions, or will a catastrophe keep resetting it over shorter intervals? This is not a matter that can be decided for all instances. Societies are not machines. History, tradition, and any number of current events can and will affect the most likely outcome in any particular instance. The pressures that are shaping a system at any given time are probably best interpreted within the framework of propensities.

8.7   Culpability

The outcome of any social action by free agents is never strictly determined. That is why, for example, we turn to judges and juries to adjudicate
the interests and/or rights of parties that come into conflict. Official judg-
ments are bound to be influenced by how those who render them perceive
agency and culpability whenever things go awry. It should be noted that
much of Western law has evolved since the Enlightenment and has been
colored strongly by the Newtonian worldview (Jaurrero-Roque 1991):
when something goes wrong, it is necessary to identify and punish the efficient agent at fault. In the large majority of cases, this is a straightforward
task. Problems arise, however, when responsibility is more diffuse—e.g.,
when nonefficient causes come into play. Our Newtonian tradition nonetheless demands that we identify guilty individuals. In American justice,
when anything goes wrong, *someone* has to pay.

Certainly, I am not advocating that many criminals be set free because
they were forced to their actions by adverse social and environmental
conditions. The very idea of social *force* is at odds with the notion of
propensity. Having just discussed the issue of free will, moreover, it
would be highly inconsistent to maintain here that individuals are
automata coerced into their deeds by their prevailing environments. We
all have a personal stake in our actions, and a corresponding individual
culpability when those actions turn malign. I am concerned here, however, with the less-frequent instances where causality exists largely at
the system-level, but tradition wrongly demands that we identify efficient agents as individuals. If a hurricane develops, we must find the
guilty butterfly!

A most poignant example of this phenomenon is one that never
appeared in a court of justice. It concerns the tragedy of the Challenger
space shuttle, which blew up several minutes into flight on January 28,
1986. Prior to this catastrophe, no American had died in space flight
(although three astronauts did perish in a fire during ground tests). This
remarkable record during the early days of manned flight owed much to
the massive efforts undertaken to insure flight safety through system
redundancy. Such preventive measures were effective, as long as the
country was both capable of and eager to support the cost of space flight
in a big, very redundant way.

With respect to the goal of getting a person to a given point in space
and back, safety measures appear as overhead. In nature, when overhead is greater than that necessary to maintain the system, it atrophies
in favor of changes that improve system efficiency (Conrad 1983), and

this trade-off usually continues until halted by some environmental perturbation. During the late seventies and early eighties, the American economy was less robust than it had been in the sixties. Since few citizens relish paying taxes, Congress was vigilant to trim expenditures whenever possible. The safety record for space flight was perfect. The recipient of decreasing relative appropriations, NASA received the clear signal to make its operations more efficient, even as its machines were growing more complex. But under constant or declining capacity, efficiency (ascendency) can increase only at the expense of reliability (overhead). In a way, a tragic outcome was inevitable. It was only a question of when and how.

In the aftermath of the tragedy a commission was formed to evaluate NASA, its operations and personnel, down to the minutest detail. It was a good search conducted in fine Newtonian fashion. During the whole investigation, however, I do not recall that the antagonism between efficiency and reliability was ever made clear to the public. Had this been done, the light of scrutiny might have been directed toward the larger system, where much of the blame seems in this case to reside.

8.8   Evolution by the Middle Class

The attempt to judge whether rights and responsibilities lie with the individual or with the community reminds us yet again of how agency may be related to scale. Reductionists would have us focus on the smallest entities and events in a system as the origins of change. The initiator of macroscopic change is a microscopic perturbation (Glansdorff and Prigogine 1970); behind every hurricane lies a butterfly. On the other hand, it appears that it is the largest and most active elements in a system that maintain its structure. They are the sources of "power" that give the system its present definition.

If we abandon the viewpoint that all macroscale behavior must be determinate, then no longer *must* we search among the smallest elements for the sole causes behind change. After all, these marginal elements possess little "momentum" of their own and become capable of initiating change *only* when the larger system is badly out of balance. On the other hand, the largest system elements can themselves precipitate change. Because they possess more potential impact, it is not necessary for the overall system to be unbalanced before their effects can be observed. These largest members, however, are usually already in stable relationships with one another, as they are what give the community its current persistent form.

It would appear, then, that mid-scale elements are the ones most likely to initiate system change. Furthermore, the entire scenario of change becomes different when it is originated by the intermediate powers. No longer *must* the major elements collapse somehow of their own weight, with the final push being provided by some minute player (although this may at times occur). More likely, change is initiated by shifts occasioned in the environments of the larger, system-defining elements by elements only moderately less ponderous in size and activity. In this scenario, change does not have to propagate upward from below, nor must it be impressed from above: its chief agents arise in the midst of the system itself.

That mid-sized elements should often play the pivotal roles in system change is formally and mathematically implicit in ascendency theory. We note that the Shannon uncertainty, the average mutual information, and the Kullback-Leibler cross-information all involve terms having the mathematical form $x \log x$. Now, the activity or size (x) of an element is reckoned as a probability, so that x takes on some value between zero and one. Furthermore, its magnitude appears as the multiplier of its own logarithm. Hence, for dominant elements with x near one, the logarithmic factor becomes vanishingly small. Conversely, the absolute magnitude of the logarithm of a small x is very large; this result is scaled, however, by the small magnitude itself. Thus, it is easy to demonstrate that as x approaches the value zero, the product $x \log x$ itself draws arbitrarily close to zero. The point is that the contribution $x \log x$ to the information indices is greatest for those elements that possess *intermediate* values of x (figure 8.1).

But what does the value of the ascendency have to do with its own change? The answer lies in a peculiar, but serendipitous, mathematical property possessed by all the key information indices. It can be rigorously demonstrated that the sensitivity of any of these sums of $x \log x$ terms with respect to a change in the magnitude of a particular flow is expressed precisely by the logarithm of that x (Casey 1992). Hence, not only do these indices express the current state of the system, but the separate terms also reveal how the system is likely to change in response to shifts in each of its members. As happened in section 7.6 in the discussion of limiting flows, analysis again points to the intermediate-sized elements as the likely loci for change.

8.9   Human Dignity

Whether the size or effect of an entity is deemed large, small, or intermediate, it is at least accorded some ontological status by virtue of this

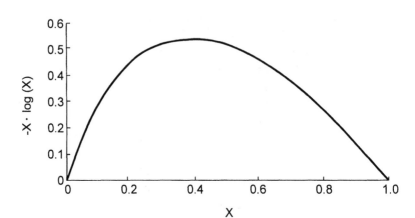

FIGURE 8.1.

A plot of the function $-x \log (x)$ vs. x, revealing how the function peaks at an intermediate value of x.

assigned magnitude. The Newtonian strictures, however, relegate certain attributes that are central to the human psyche, such as free will and self-identity, to the inferior status of epiphenomena—things that only appear to exist. Many regard this demotion as an indignity that Newtonian thinkers have attempted to force upon humankind—that people are but machines, *absolutely* subservient to physical and chemical nature and to a history that in turn is patterned by the physical world—and their angst is only heightened by the large measure of truth in this view. Daily we read of new discoveries in genetics, ontogeny, and psychology that illumine particular mechanical aspects of our being. Any number of movements in the arts and humanities arose as a conscious or semiotic rebellion against the proposition of *Homo sapiens* as automaton: art and poetry as sheer expressions of *will* flung their gauntlets in the face of physical reductionism. Defiance, of course, was hardly one-sided, as scientists deflated one belief after another with the bravado of someone in the throes of discovery.

The postmodern perspective evolving during our times, however, represents a clear break with the direction of the past three hundred years, just as the Newtonian revolution was itself a profound shift away from premodern thought. No one wishes to return to the past; history cannot and should not repeat itself. Nevertheless, many will find solace in knowing that the new outlook mitigates much of the conflict between science and society that has characterized the time since Newton's

Principia (Goerner 1993). Science as philosophy is maturing, and with maturity comes the realization of self-limitations, as Thomas Kuhn, Paul Feyerabend, and countless others argue against the privileged position of science vis-à-vis other, more subjective human endeavors. In spite of almost daily revelations from genetic research, the goal of physical reductionism seems to grow ever less feasible.

The prospect now is for a "kindlier, gentler" science practiced by investigators fully aware that they are part and parcel of the systems they observe (Matsuno 1989). Measures of living systems will still be taken, and predictions of behavior will still be made. But the probabilistic nature of those predictions will be viewed, not as a weakness (as is the current attitude toward the "soft" sciences), but as explicit recognition that residual freedom in nature is absolutely essential to the persistence, creation, and health of living systems.

8.10   Science and Religion

This would be a hopeful note upon which to end this essay. One other topic bearing upon the relationship between science and society remains, however, to be addressed. The roots that fed Newtonianism did not seem to emanate from the arts and humanities, but from religion and politics. Or rather, to put a finer point on it, from religion *in* politics. I speculated earlier that the philosophical target of the revolution was late scholasticism, as typified by Aquinas's assertion of God as final cause. A more immediate political motivation was likely a widespread resentment against what was perceived as the undue influence of clerics upon all levels of society. One effective way of undermining clericalism was to attack the very beliefs upon which clerical authority rested, and that avenue was vigorously pursued. As a result, part of our Newtonian inheritance is the notion that religion and science are not just mutually exclusive, but also fundamentally opposed.

Oddly enough, however, one of the early accomplishments of the Enlightenment appears to have been its power to resolve church-state and church-individual conflicts on a political level. (Here the American model of "separation of church and state" looms large as an example.) That resolution was that a secular democracy should not include any component of any organized religion as a constitutive element; at the same time, however, a government may not curtail or inhibit the formation of religious groups and beliefs. This accommodation effectively achieved what many of Newton's contemporaries sought, namely, freedom of thought and speech. Furthermore, it was accomplished by employing a form of "atomism." Imperfect though its implementation sometimes may be, no one living under this accord is *compelled* to espouse a dogmatic view of the world. Should any individual member of a religious group develop irreconcilable differences with the rest of his or her community, that individual is guaranteed the right to separate from the group without suffering civil recriminations. Furthermore, everyone is free to not join any religious group or profess any religious belief whatsoever.

While it may seem that organized religion has lost much of its power through this compromise, it is hardly regarded as a defeat by most contemporary religionists. Most Western believers have come to appreciate the power afforded by working within a group that shares a consensus of beliefs. The idea that God would compel belief is probably as foreign to today's believers as is the notion that physical nature fully determines the actions of an individual. Furthermore, the concept of separating secular and religious powers has proved a useful tool to some religious groups existing under totalitarian ideologies (e.g., Weigel 1992). Finally, secular power can still be exercised by religious communities, who under the contract are free to exert moral suasion, as did the prophets of the "Book" religions.

In short, the political and social conditions that helped motivate the Newtonian revolution have long ago faded from the scene. Yet as long as the Newtonian paradigm still represents the broad consensus of scientific thought, an ostensible antagonism will persist between science and religion. Might not the evolving ecological worldview ameliorate at least the magnitude of this conflict? While my own answer is a definite, but qualified, "yes," I would hasten immediately to the disclaimers. We should remain suspicious, for example, of anyone who attempts to invoke science or logic to either "prove" or "disprove" propositions that necessarily remain matters of faith, be they of the scientific or the religious variety. One cannot, for example, prove the existence of God. For to make a watertight case would immediately make obligate the belief in the Supreme Being—but a forced believer is little better than an automaton, and forced belief hardly seems worthy of the God most people worship.

The lessons learned from the Enlightenment resolution of tensions over religion and politics need to be applied to the conceptual dialogue as well. That the ethical pursuit of science should remain autonomous of religious belief is an outgrowth of the scientific revolution that virtually no one contests. On the other hand, the creativity, diversity, plasticity, and indeterminacy of human nature are quite sufficient to guarantee that religion will never be wholly displaced or subsumed by science. The two domains of human knowledge will always maintain significant autonomy from each other. But is there any overlap between them, and, if so, does the ecological perspective in any way illumine a common ground (Ferré 1982; Ulanowicz 1995a)?

Some point to a possible connection between ecology and religion via the theory known as the "Gaia hypothesis." James Lovelock (1979) described certain biogeophysical feedback processes thought to regulate global conditions in such a way that they always maintain an environment suitable for life. For example, the plants that cover the various surfaces of the earth can live only within a limited range of temperatures. According to Lovelock, changing flora modulate the albedo (reflective power) of the earth over time in such a way as to maintain the global temperature within the window of viability.

It is understandable that Lovelock should describe global feedback in metaphysical terms (especially if by "metaphysical" one means "non-Newtonian"). He went further, however, and named the phenomenon "Gaia" after the Greek goddess of the earth. Such numinous overtones have led some to deify Gaia as an object of reverence, due in return for her providence. Gaia comes close to fulfilling the desire of many New Age advocates for a science that subsumes religious belief. Others, such as Thomas Berry (1987), point toward a new religion of "Ecology Arising." But most ecologists eschew the mantle of priesthood in the New Order, despite the fact that they already preach the necessity to respect and guard the life-sustaining processes on earth. For many it is probably an unconscious faith in Newtonianism that leads them to reject the exaggerated forms of Gaia. What they seem to disdain most is the resemblance that Gaia bears to Clements's earlier identification of the ecosystem as a superorganism.

Most systems ecologists concede that the capacity for some regulation of self and environment is a prerequisite if living systems are to survive for any appreciable length of time. Organic homeostasis alone, however, even when scaled by the enormity of the biosphere and the duration of its processes, remains insufficient to establish the global ecosystem as an entity superior to an organism or a species. As Depew and Weber (1994) have said about Clements, he had it backward: Ecosystems are not superorganisms; organisms are superecosystems. The same inversion of status is cited by the religionist, who sees in the Babylonian who prayed to the sun that warmed and sustained him similarities to the wolf that bays at the moon. The fact that the sun is enormously larger than the Babylonian and endures for epochs before and beyond the transitory petitioner does not mean that the sun or the moon can begin to compare with the Babylonian, or even any higher animal, as an entity of complexity and adaptability.

The ecological perspective built around the ideas of Popper and Aristotle would seem, in my opinion, to provide a metaphor that might be more effective at ameliorating the conflict between science and religion without requiring that either be subsumed by the other, or that reconciliation be sought in some less-fecund hybrid. As I have outlined, ecosystem dynamics resemble a dialectic between agencies that create structure, on the one hand, and the inexorable tendency for those structures to fall apart, on the other. One could regard religion and science as leaning against each other in much the same way.

At one level there will always remain a basic tension between the two modes of thought. Science will continue to prune religion of that which is magical or superstitious. The religious community will persist in articulating virtues that, in the end, delimit the practice of science. Inescapably, such interaction will continue to cause discomfort to any and all participants in the dialogue. On a more positive note, we have seen how a dialectic never occurs on a single hierarchical level: just as a direct competition between two populations can become synergistic at the higher level of the ecosystem, so the interaction between science and religion can, at the next level, prove mutualistic. For the believer, faith in the transcendent can sustain the arduous pursuit of scientific knowledge far beyond the motivating powers of social and financial rewards. Likewise, the very faith that the nonbeliever places in secular principles could predispose him or her toward the (too often neglected) activity of scientific induction.

Looking in the other direction, the rise of a critical spirit, so fundamental to the pursuit of science, exacts for many who would employ it "a more personal and explicit adherence to faith. As a result many persons are achieving a more vivid sense of God" (Vatican II 1965:7). It is the pursuit of this form of knowledge of the divine that motivates many of the Bible study groups that now proliferate in Western churches. Furthermore, it sometimes happens that new avenues are opened for religious thought by parallel discoveries in secular science. For example, Ilya Prigogine and Isabel Stengers (1984) were eloquent in describing how the theory of dissipative structures frees the mind from the bondage imposed by Newtonian determinism. Humans are no longer perceived as automata forced to act in rigid synchrony with an all-pervasive Deist clockwork; rather, we now are free to pursue a new "dialogue with nature." Believers see in this freedom the avenue by which they enter into their dialogue with God, for, in a more loosely coupled world, the ways in which God can respond no longer seem so improbable. Thus, it is probably no accident that the new thread of "process theology" bears a marked similarity to the program by which Prigogine suggests we pursue our new dialogue with nature (Haught 1984).

8.11   New Century, New Millennium, Fresh Outlook

What astounding changes in our perspective on what is possible (and natural) can result as soon as we reject the stricture that all causes must be limited to the material or mechanical! This is not to belittle the genius of those who first proposed the constraint three hundred years ago: such a narrow focus has proved spectacularly expedient in effectively and quantitatively describing phenomena that occur far from our immediate experience. Events not readily accessible to our immediate senses do seem to be simpler in essence than those that mire our daily lives. It does not follow, however, that what is efficacious in distant realms provides the only legitimate description of more immediate affairs. It would appear that Occam's razor truly is a double-edged blade: in recent centuries its posterior edge has cut from our view perfectly natural and quantitative descriptions of life as we most immediately encounter it. It is dogma, pure and simple, that would have us believe that formal agency and directed behavior are but epiphenomena, and not legitimate components of the natural world.

It was Arne Naess (1988) who introduced the notion of "deep ecology" to refer to the implicit penetration of the discipline into other fields, and possibly even into the human condition itself. Doubtless, many readers will choose not to follow its traces back as far as I have explored them here. Each reader should choose what seems plausible and ignore the rest. One point, however, I sincerely hope will be accepted before all others: Ecology should not be deemed a sick science, an immature science, or a pseudo-science simply because it cannot readily be twisted to conform to descriptions that pertain to the limits of observation. Ecology deals with a world vastly richer than what can be extrapolated from the fringes of existence. As we begin to develop the tools that will help us describe and decode this richness, we acquire a new respect for this "orphan" of science. The long-neglected stepchild is about to take center stage. Ecology is on its way to becoming the ascendant perspective of the next century!

REFERENCES

Abrams, P. 1996. Dynamics and interactions in food webs with adaptive foragers. Pp. 113–121. In G. Polis and K. Winemiller, eds., *Food Webs: Integration of Patterns and Dynamics.* Chapman Hall, NY.

Ahl, V. and T. F. H. Allen. Forthcoming. *Hierarchy Theory: A Vision, Vocabulary, and Epistemology.* Columbia University Press, NY.

Allen, T. F. H. and T. W. Hoekstra. 1992. *Toward a Unified Ecology.* Columbia University Press, NY. 384 pp.

Allen, T. F. H. and T. B. Starr. 1982. *Hierarchy.* University of Chicago Press, Chicago. 310 pp.

Atlan, H. 1974. On a formal definition of organization. *J. theor. Biol.* 45:295–304.

Ayres, R. U. 1994. *Information, Entropy and Progress: A New Evolutionary Paradigm.* AIP Press, NY. 301 pp.

Baird, D. and R. Ulanowicz. 1989. The seasonal dynamics of the Chesapeake Bay ecosystem. *Ecol. Monogr.* 59:329–364.

Baird, D., R. E. Ulanowicz, and W. R. Boynton. 1995. Seasonal nitrogen dynamics in the Chesapeake Bay: A network approach. *Estuar. Coast. Shelf Sci.* 41:137–162.

Bateson, G. 1979. *Mind and Nature: A Necessary Unity.* Dutton, NY. 238 pp.

Berry, T. 1987. The dream of the earth: Our way into the future. *Cross Currents* 37:200–215.

Bertness, M. D. and R. Callaway. 1994. Positive interactions in communities. *Trends Ecol. Evol.* 9:191–193.

Bertness, M. D. and S. D. Hacker. 1994. Physical stress and positive associations among marsh plants. *Am. Nat.* 144:363–372.

Black, J. 1806. *Lectures on the Elements of Chemistry Delivered at the University of Edinburgh.* M. Carey, Philadelphia. 196 pp.

Bossel, H. 1987. Viability and sustainability: Matching development goals to resource constraints. *Futures* 19:114–128.

Bosserman, R. W. 1979. The hierarchical integrity of *Utricularia*-periphyton microecosystems. Ph.D. diss., University of Georgia. 266 leaves.

Bosworth, K. W. and R. E. Ulanowicz. 1991. The possible role of ascendency in neural networks. (Abstract) American Mathematical Society, Southeastern Sectional Meeting, March 22, Tampa, FL.

Brooks, D. R. and E. O. Wiley. 1986. *Evolution as Entropy: Toward a Unified Theory of Biology.* University of Chicago Press, Chicago. 335 pp.

Campbell, D. T. 1974. "Downward causation" in hierarchically organized biological systems. Pp. 179–186. In F. J. Ayala and T. Dobzhansky, eds., *Studies in the Philosophy of Biology.* University of California Press, Berkeley.

Carnot, S. 1824. *Reflections on the Motive Power of Heat* (translated 1943). ASME, NY. 107 pp.

Casey, A. 1992. Influence theory for trophic networks. Master's thesis, University of Maryland. 80 pp.

Casti, J. 1989. Newton, Aristotle, and the Modeling of Living Systems. Pp. 47–89. In J. Casti and A. Karlqvist, eds., *Newton to Aristotle.* Birkhaeuser, NY.

———. 1992. That's life?—Yes, No, Maybe. (Pre-print, no pages given.) In *Proceedings of the Frontiers of Life Workshop, Blois, France, October 1991.*

Cesarman, E. 1996. *Thermodynamics of the Heart.* Robles, Mexico City. 191 pp.

Chapman, S. and T. G. Cowling. 1961. *The Mathematical Theory of Nonuniform Gases.* Cambridge University Press, Cambridge. 431 pp.

Cheslak, E. F. and V. A. Lamarra. 1981. The residence time of energy as a measure of ecological organization. Pp. 591–600. In W. J. Mitsch, R. W. Bosserman, and J. M. Klopatek, eds., *Energy and Ecological Modelling.* Elsevier, NY.

Christensen, V. 1994. On the behavior of some proposed goal functions for ecosystem development. *Ecol. Model.* 75/76:37–49.

Christian, R. R., J. N. Boyer, D. W. Stanley, and W. M. Rizzo. 1992. Network analysis of nitrogen cycling in an estuary. Pp. 217–247. In C. J. Hurst, ed., *Modeling the Metabolic and Physiology Activities of Microorganisms.* Wiley, NY.

Christian, R. R., E. Fores, F. Comin, P. Viaroli, M. Naldi, and I. Ferrari. 1996. Nitrogen cycling networks of coastal ecosystems: Influence of trophic status and primary producer form. *Ecol. Model.* 87:111–129.

Clements, F. E. 1916. *Plant Succession: An Analysis of the Development of Vegetation.* Carnegie Institution of Washington, Washington, D.C. 340 pp.

Cohen, J. E. 1971. Mathematics as metaphor. *Science* 172:674–675.

Conrad, M. 1983. *Adaptability: The Significance of Variability from Molecule to Ecosystem.* Plenum Press, NY. 383 pp.

Costanza, R. 1992. Toward an operational definition of ecosystem health. Pp. 239–256. In R. Costanza, B. G. Norton, and B. D. Haskell, eds., *Ecosystem Health: New Goals for Environmental Management.* Island Press, Washington, D.C. 269 pp.

Costanza, R., H. E. Daly, and J. A. Bartholomew. 1991. Goals, agenda, and policy recommendations for ecological economics. Pp. 1–12. In R. Costanza, ed., *Ecological Economics: The Science and Management of Sustainability.* Columbia University Press, NY.

Crick, F. H. 1982. *Life Itself: Its Origin and Nature.* Simon and Schuster, NY. 192 pp.

Dawkins, R. 1976. *The Selfish Gene.* Oxford University Press, NY. 224 pp.

DeAngelis, D. L. and L. J. Gross, eds., 1992. *Individual-based Models and Approaches in Ecology: Populations, Communities, and Ecosystems.* Chapman and Hall, NY. 525 pp.

DeAngelis, D. L., W. M. Post, and C. C. Travis. 1986. *Positive Feedback in Natural Systems.* Springer-Verlag, NY. 290 pp.

Depew, D. J. and B. H. Weber. 1988. Consequences of nonequilibrium thermodynamics for the Darwinian tradition. Pp. 317–354. In B. Weber, D. Depew, and J. Smith., eds., *Entropy, Information, and Evolution: New Perspectives on Physical and Biological Evolution.* MIT Press, Cambridge, MA.

———. 1994. *Darwinism Evolving: Systems Dynamics and the Genealogy of Natural Selection.* MIT Press, Cambridge, MA. 588 pp.

Dicke, R. H. and J. P. Wittke. 1960. *Introduction to Quantum Mechanics.* Addison-Wesley, Reading, MA. 369 pp.

Edmondson, W. T. 1970. Phosphorus, nitrogen, and algae in Lake Washington after diversion of sewage. *Science* 169:690–691.

Edwards, C. J. and H. A. Regier. 1990. *An Ecosystem Approach to the Integrity of the Great Lakes in Turbulent Times.* Great Lakes Fishery Commission, Ann Arbor, MI. 299 pp.

Elliot, J. E. 1980. Marx and Schumpeter on capitalism's creative destruction: A comparative restatement. *Quart. J. Econ.* 95:46–58.

Engelberg, J. and L. L. Boyarsky. 1979. The noncybernetic nature of ecosystems. *Am. Nat.* 114:317–324.

Evans, R. B. 1969. A proof that essergy is the only consistent measure of potential work. Ph.D. diss., Dartmouth College. 129 pp.

Faber, M. and R. Manstetten. 1991. Ihr werdet sein wie Gott. *Neue Zuericher Zeitung*, 31 March 1991, 14:28–29.

Ferguson, M. 1980. *The Aquarian Conspiracy: Personal and Social Transformation in the 1980s*. St. Martin's, NY. 448 pp.

Ferré, F. 1982. Religious world modeling and postmodern science. *J. Relig.* 62:261–271.

Field, J. G., C. L. Moloney, and C. G. Attwood. 1989. Network analysis of simulated succession after an upwelling event. Pp. 132–158. In F. W. Wulff, J. G. Field, and K. H. Mann, eds., *Network Analysis in Marine Ecology: Methods and Applications*. Springer-Verlag, Berlin.

Finn, J. T. 1976. Measures of ecosystem structure and function derived from analysis of flows. *J. theor. Biol.* 56:363–380.

Fisher, R. A. 1930. *The Genetical Theory of Natural Selection*. Oxford University Press, Oxford. 272 pp.

Fleming, D. M., D. L. DeAngelis, L. J. Gross, R. E. Ulanowicz, W. F. Wolff, W. F. Loftus, and M. A. Huston. 1995. *ATLSS: Across-Trophic Level System Simulation for the Freshwater Wetlands of the Everglades and Big Cypress Swamp*. National Biological Service, South Florida/Caribbean Field Laboratory, Homestead, FL. 76 pp.

Ford, K. W., G. I. Rochlin, and R. H. Socolow. 1975. *Efficient Use of Energy*. American Institute of Physics, NY. 304 pp.

Forrest, S. 1991. *Emergent Computation: Self-Organizing, Collective, and Cooperative Phenomena in Natural and Artificial Computing Networks*. MIT Press, Cambridge, MA. 452 pp.

Franklin, J. 1983. *Not Quite a Miracle: Brain Surgeons and Their Patients on the Frontier of Medicine*. Doubleday, Garden City, NY. 274 pp.

Gaggiolo, R. A. 1980. *Thermodynamics: Second Law Analysis*. American Chemical Society, Washington, D.C. 301 pp.

Garrett, L. 1993. AIDS researchers seeing their quarry in new light. *Baltimore Sun*, September 7, pp. 1, 7.

Glansdorff, P. and I. Prigogine. 1970. Non-equilibrium stability theory. *Physica* 46:344–366.

Gleason, H. A. 1917. The structure and development of the plant association. *Bull. Torrey Bot. Club* 44:463–481.

Goerner, S. J. 1993. *Chaos and the Evolving Ecological Universe: A Study in the Science and Human Implications of a New World Hypothesis*. Gordon and Breach, NY. 255 pp.

Golley, F. B. 1974. Structural and functional properties as they influence ecosystem stability. Pp. 97–102. In A. J. Cave, ed., *Proceedings of the First International Congress of Ecology*. Centre for Agricultural Publishing and Documentation, Wageningen, Netherlands.

——. 1993. *A History of the Ecosystem Concept in Ecology.* Yale University Press, New Haven, CT. 254 pp.

Golub, G. H. and C. F. Van Loan. 1983. *Matrix Computations.* Johns Hopkins University Press, Baltimore. 476 pp.

Gore, A., Jr. 1992. *Earth in the Balance: Ecology and the Human Spirit.* Houghton Mifflin, Boston. 407 pp.

Hagen, J. B. 1992. *An Entangled Bank: The Origins of Ecosystem Ecology.* Rutgers University Press, New Brunswick, NJ. 245 pp.

Hannon, B. 1973. The structure of ecosystems. *J. theor. Biol.* 41:535–546.

——. 1979. Total energy costs in ecosystems. *J. theor. Biol.* 80:271–293.

Hannon, B., R. Costanza, and R. E. Ulanowicz. 1991. A general accounting framework for ecological systems: A functional taxonomy for connectivist ecology. *Theor. Popul. Biol.* 40:78–104.

Harner, M. J. 1970. Population pressure and the social evolution of agriculturalists. *Southwestern J. Anthro.* 26:67–86.

Hatsopoulos, G. and J. Keenan. 1965. *Principles of General Thermodynamics.* Wiley, NY. 788 pp.

Haught, J. F. 1984. *The Cosmic Adventure: Science, Religion, and the Quest for Purpose.* Paulist Press, NY. 184 pp.

Hevert, H. and S. Hevert. 1980. Second law analysis: An alternative indicator of system efficiency. *Energy — Int. J.* 5:865–873.

Hirata, H. and R. E. Ulanowicz. 1984. Information theoretical analysis of ecological networks. *Int. J. Syst. Sci.* 15:261–270.

Hodge, M. J. S. 1992. Biology and philosophy (including ideology): A study of Fisher and Wright. Pp. 231–293. In S. Sarker, ed., *The Founders of Evolutionary Genetics.* Kluwer, Dordrecht.

Hogg, T. and B. A. Huberman. 1991. Controlling chaos in distributed systems. *IEEE Trans. Syst. Man Cybernet.* 21 (6): 1325–1332.

Holling, C. S. 1986. The resilience of terrestrial ecosystems: Local surprise and global change. Pp. 292–317. In W. C. CLark and R. E. Munn, eds., *Sustainable Development of the Biosphere.* Cambridge University Press, Cambridge.

——. 1992. Cross-scale morphology, geometry, and dynamics of ecosystems. *Ecol. Monogr.* 62 (4): 447–502.

Homer, M., W. M. Kemp, and H. McKellar. 1976. Trophic analysis of an estuarine ecosystem: Salt marsh–tidal creek system near Crystal River, Florida. Unpublished manuscript, Department of Environmental Engineering, University of Florida, Gainesville.

Hopfield, J. J. and D. W. Tank. 1986. Computing with neural circuits: A model. *Science* 233:625–633.

Jaurrero-Roque, A. 1991. Fail-safe versus safe-fail: Suggestions toward an evolutionary model of justice. *Texas Law Rev.* 69:1745–1777.

Johnson, L. 1990. The thermodynamics of ecosystems. Pp. 2–46. In O. Hutzinger, ed., *The Handbook of Environmental Chemistry*, vol. 1, *The Natural Environment and the Biogeochemical Cycles*. Springer-Verlag, Heidelberg.

Jorgensen, S. E. 1992. *Integration of Ecosystem Theories: A Pattern*. Kluwer, Dordrecht. 383 pp.

Jorgensen, S. E. and H. Mejer. 1979. A holistic approach to ecological modelling. *Ecol. Model.* 7:169–189.

Kauffman, S. A. 1991. Antichaos and adaptation. *Sci. Am.* 265:78–84.

Keenan, J. H. 1951. Availability and irreversibility in thermodynamics. *Br. J. Appl. Phys.* 2:183–192.

Knight, S. E. and T. M. Frost. 1991. Bladder control in *Utricularia macrorhiza*: Lake-specific variation in plant investment in carnivory. *Ecology* 72 (2): 728–734.

Kuhn, T. S. 1970. *The Structure of Scientific Revolutions*. 2d ed. University of Chicago Press, Chicago. 210 pp.

Langton, C. G. 1992. Life at the edge of chaos. In C. G. Langton et al., eds., *Artifical Life II: The Proceedings of the Workshop on Artificial Life Held in 1990 in Santa Fe, New Mexico*. Addison-Wesley, Reading, MA. 854 pp.

Laplace, P. S. 1814. *A Philosophical Essay on Probabilities*. Dover, NY. 196 pp.

Lemons, J. and L. Westra. 1995. *Perspectives on Ecological Integrity*. Kluwer, Boston. 279 pp.

Leslie, P. H. 1945. The use of matrices in certain population mathematics. *Biometrika* 33:183–212.

Lewin, R. 1984. Why is development so illogical? *Science* 224:1327–1329.

Liebig, J. 1854. *Chemistry in Its Application to Agriculture and Physiology*. Taylor and Walton, London. 401 pp.

Lindeman, R. L. 1942. The trophic-dynamic aspect of ecology. *Ecology* 23:399–418.

Lotka, A. J. 1922. Contribution to the energetics of evolution. *Proc. Natl. Acad. Sci. USA* 8:147–150.

Lovelock, J. E. 1979. *Gaia: A New Look at Life on Earth*. Oxford University Press, NY. 157 pp.

MacArthur, R. 1955. Fluctuations of animal populations, and a measure of community stability. *Ecology* 36:533–536.

MacArthur, R. H. and E. O. Wilson. 1967. *Theory of Island Biogeography*. Princeton University Press, Princeton, NJ. 203 pp.

Mageau, M. T., R. Costanza, and R. E. Ulanowicz. 1995. The development and testing of a quantitative assessment of ecosystem health. *Ecosyst. Health* 1 (4): 201–213.

Manuel, F. E. 1968. *A Portrait of Isaac Newton*. Harvard University Press, Cambridge, MA. 478 pp.

Margalef, R. 1968. *Perspectives in Ecological Theory*. University of Chicago Press, Chicago. 111 pp.

Matsuno, K. 1989. *Protobiology: Physical Basis of Biology*. CRC Press, Boca Raton, FL. 225 pp.

——. 1993. Being free from *ceteris paribus*: A vehicle for founding physics on biology rather than the other way around. *Appl. Math. Compt.* 56:261–279.

May, R. M. 1973. *Stability and Complexity in Model Ecosystems*. Princeton University Press, Princeton, NJ. 235 pp.

Mayr, E. 1978. Evolution. *Science* 239:46–55.

——. 1992. The idea of teleology. *J. Hist. Ideas* 53 (1): 117–177.

Mickulecky, D. C. 1985. Network thermodynamics in biology and ecology: An introduction. Pp. 163–175. In R. E. Ulanowicz and T. Platt, eds., *Ecosystem Theory for Biological Oceanography*. Canadian Bulletin of Fisheries and Aquatic Sciences 213, Ottawa.

Moore, W. J. 1962. *Physical Chemistry*. Prentice-Hall, Englewood Cliffs, NJ. 844 pp.

Morowitz, H. J. 1986. Entropy and nonsense. *Biol. Phil.* 1:473–476.

Myers, R. L. 1990. Scrub and high pine. Pp. 150–193. In R. L. Myers and J. J. Ewel, eds., *Ecosystems of Florida*. University of Central Florida Press, Orlando.

Naess, A. 1988. Deep ecology and ultimate premises. *Ecologist* 18:128–131.

Norton, B. G. and R. E. Ulanowicz. 1992. Scale and biodiversity policy: A hierarchical approach. *Ambio* 21 (3): 244–249.

Odum, E. P. 1953. *Fundamentals of Ecology*. Saunders, Philadelphia. 384 pp.

——. 1959. *Fundamentals of Ecology*. 2d ed. Saunders, Philadelphia. 546 pp.

——. 1969. The strategy of ecosystem development. *Science* 164:262–270.

Odum, H. T. 1960. Ecological potential and analogue circuits for the ecosystem. *Am. Sci.* 48:1–8.

——. 1971. *Environment, Power, and Society*. Wiley, NY. 331 pp.

——. 1982. Pulsing, power, and hierarchy. Pp. 33–59. In W. J. Mitsch, R. K. Ragade, R. W. Bosserman, and J. A. Dinnon, Jr., eds., *Energetics and Systems*. Ann Arbor Science, Butterworth Group, Ann Arbor, MI.

——. 1988. Self-organization, transformity, and information. *Science* 242:1132–1139.

Odum, H. T. and R. C. Pinkerton. 1955. Time's speed regulator: The optimum efficiency for maximum power output in physical and biological systems. *Am. Sci.* 43:331–343.

Onsager, L. 1931. Reciprocal relations in irreversible processes. *Phys. Rev. A* 37:405–426.

Pahl-Wostl, C. 1992. Information theoretical analysis of functional temporal and spatial organization in flow networks. *Math. Comput. Model.* 16 (3): 35–52.

———. 1995. *The Dynamic Nature of Ecosystems: Chaos and Order Entwined.* Wiley, NY. 267 pp.

Patten, B. C. and M. Higashi. 1991. Network ecology: Indirect determination of the life-environment relationship in ecosystems. Pp. 288–351. In M. Higashi and T. Platt, eds., *Theoretical Ecosystem Ecology: The Network Perspective.* Cambridge University Press, London.

Peters, R. H. 1993. *A Critique for Ecology.* Cambridge University Press, Cambridge. 366 pp.

Pimm, S. L. 1982. *Food Webs.* Chapman and Hall, London. 219 pp.

Platt, T. 1985. Structure of the marine ecosystem: Its allometric basis. Pp. 55–64. In R. E. Ulanowicz and T. Platt, eds., *Ecosystem Theory for Biological Oceanography.* Canadian Bulletin of Fisheries and Aquatic Sciences 213, Ottawa.

Platt, T., K. H. Mann, and R. E. Ulanowicz, eds. 1981. *Mathematical Models in Biological Oceanography.* UNESCO Press, Paris. 157 pp.

Popper, K. R. 1982. *The Open Universe: An Argument for Indeterminism.* Rowman and Littlefield, Totowa, NJ. 185 pp.

———. 1990. *A World of Propensities.* Thoemmes, Bristol. 51 pp.

Porter, T. M. 1986. *The Rise of Statistical Thinking, 1820–1900.* Princeton University Press, Princeton, NJ. 333 pp.

Prigogine, I. 1945. Modération et transformations irréversibles des systèmes ouverts. *Bull. Class. Sci. Acad. R. Belg.,* 5th ser., 31:600–606.

———. 1967. *Introduction to the Thermodynamics of Irreversible Processes.* 3d ed. Interscience, NY. 147 pp.

Prigogine, I. and I. Stengers. 1984. *Order out of Chaos: Man's New Dialogue with Nature.* Bantam, NY. 349 pp.

Reynolds, W. C. and H. C. Perkins. 1977. *Engineering Thermodynamics.* McGraw-Hill, NY. 585 pp.

Rosen, R. 1985. Information and complexity. Pp. 221–233. In R. E. Ulanowicz and T. Platt, eds., *Ecosystem Theory for Biological Oceanography.* Canadian Bulletin of Fisheries and Aquatic Sciences 213, Ottawa.

———. 1991. *Life Itself: A Comprehensive Inquiry into the Nature, Origin, and Foundation of Life.* Columbia University Press, NY. 285 pp.

Rossler, O. E. 1987. Ecophysics. Pp. 24–46. In J. L. Casti and A. Karlqvist, eds., *Real Brains, Artificial Minds.* North-Holland, Amsterdam.

Rutledge, R. W., B. L. Basorre, and R. J. Mulholland. 1976. Ecological stability: An information theory viewpoint. *J. theor. Biol.* 57:355–371.

Ryther, J. H. and W. M. Dunstan. 1971. Nitrogen, phosphorus, and eutrophication in the coastal marine environment. *Science* 171:1008–1013.

Salthe, S. N. 1985. *Evolving Hierarchical Systems: Their Structure and Representation.* Columbia University Press, NY. 343 pp.

——. 1993. *Development and Evolution: Complexity and Change in Biology.* MIT Press, Cambridge, MA. 257 pp.

Schneider, E. D. and J. J. Kay. 1994a. Complexity and thermodynamics: Towards a new ecology. *Futures* 24 (6): 626–647.

——. 1994b. Life as a manifestation of the second law of thermodynamics. *Math. Comput. Model.* 19:25–48.

Schneider, T. D. 1991a. Theory of molecular machines. I. Channel capacity of molecular machines. *J. theor. Biol.* 148:83–123.

——. 1991b. Theory of molecular machines. II. Energy dissipation from molecular machines. *J. theor. Biol.* 148:125–137.

Schweber, S. S. 1979. Essay review: The young Darwin. *J. Hist. Biol.* 12:175–192.

Service, E. R. 1960. The law of evolutionary potential. Pp. 93–122. In M. D. Sahlins and E. R. Service, eds., *Evolution and Culture.* University of Michigan Press, Ann Arbor. 131 pp.

Shannon, C. E. 1948. A mathematical theory of communication. *Bell System Tech. J.* 27:379–423.

Simberloff, D. 1980. A succession of paradigms in ecology: Essentialism to materialism and probabilism. *Synthese* 43:3–39.

Simon, H. A. 1956. Rational choice and the structure of the environment. *Psych. Rev.* 63 (2): 129–138.

Swenson, R. 1989a. Emergent attractors and the law of maximum entropy production: Foundations to a theory of general evolution. *Syst. Res.* 6:187–197.

——. 1989b. Emergent evolution and the global attractor: The evolutionary epistemology of entropy production maximization. Pp. 46–53. In *Proceedings of 33rd Annual Meeting ISSS, 2–7 July, Edinburgh, Scotland.* ISSS, Edinburgh.

Swimme, B. and T. Berry. 1992. *The Universe Story.* Harper-Collins, San Francisco. 305 pp.

Tainter, J. A. 1988. *The Collapse of Complex Societies.* Cambridge University Press, NY. 250 pp.

Tansley, A. G. 1935. The use and abuse of vegetational concepts and terms. *Ecology* 16:284–307.

Tolstoy, L. 1911. *War and Peace,* vol. 3. Dutton, NY.

Transeau, E. N. 1926. The accumulation of energy by plants. *Ohio Journal of Science* 26:1–10.

Tribus, M. 1961. The formalism of statistical mechanics. Pp. 69–88. In M. Tribus, ed., *Thermostatics and Thermodynamics.* Van Nostrand, Princeton, NJ.

Tribus, M. and E. C. McIrvine. 1971. Energy and information. *Sci. Am.* 225:179–188.

Ulanowicz, R. E. 1972. Mass and energy flow in closed ecosystems. *J. theor. Biol.* 34:239–253.

——. 1979. Prediction, chaos, and ecological perspective. Pp. 107–117. In E. A. Halfon, ed., *Theoretical Systems Ecology.* Academic Press, NY.

——. 1980. An hypothesis on the development of natural communities. *J. theor. Biol.* 85:223–245.

——. 1983. Identifying the structure of cycling in ecosystems. *Math. Biosci.* 65:219–237.

——. 1984. Community measures of marine food networks and their possible applications. Pp. 23–47. In M. J. R. Fasham, ed., *Flows of Energy and Materials in Marine Ecosystems.* Plenum, NY.

——. 1986a. *Growth and Development: Ecosystems Phenomenology.* Springer-Verlag, NY. 203 pp.

——. 1986b. A phenomenological perspective of ecological development. Pp. 73–81. In T. M. Poston and R. Purdy, eds., *Aquatic Toxicology and Environmental Fate*, vol. 9, *ASTM Spec. Tech. Pub. 921.* American Society for Testing and Materials, Philadelphia.

——. 1989. A phenomenology of evolving networks. *Syst. Res.* 6:209–217.

——. 1991. Contributory values of ecosystem resources. Pp. 253–268. In R. Costanza, ed., *Ecological Economics: The Science and Management of Sustainability.* Columbia University Press, NY.

——. 1992. Ecosystem health and trophic flow networks. Pp. 190–205. In R. Costanza, B. G. Norton, and B. D. Haskell, eds., *Ecosystem Health: New Goals for Environmental Management.* Island Press, Washington, D.C.

——. 1993. Perspectives: Oecologia ex machina? *Ecomod.* 11 (2): 1, 9.

——. 1995a. Beyond the material and the mechanical: Occam's razor is a double-edged blade. *Zygon* 30 (2): 249–266.

——. 1995b. Ecosystem integrity: A causal necessity. Pp. 77–87. In J. Lemons and L. Westra., eds., *Perspectives on Implementing Ecological Integrity.* Kluwer, Dordrecht.

——. 1995c. Ecosystem trophic foundations: Lindeman Exonerata. Pp. 549–560. In B. C. Patten and S. E. Jorgensen, eds., *Complex Ecology: The Part-Whole Relation in Ecosystems.* Prentice-Hall, NY.

——. 1995d. Utricularia's secret: The advantage of positive feedback in oligotrophic environments. *Ecol. Model.* 79:49–57.

——. 1996. The propensities of evolving systems. Pp. 217–233. In E. L. Khalil and K. E. Boulding, eds., *Evolution, Order and Complexity.* Routledge, London. 276 pp.

——. Forthcoming. Boundaries on the complexity of evolving networks: A window of vitality. In A. Franchini, ed., *Biological Models: Proceedings of the 1992 Summer School on Environmental Dynamics.* Istituto Veneto di Scienze, Lettere ed Arti, Venice.

Ulanowicz, R. E. and L. G. Abarca. Forthcoming. An informational synthesis of ecosystem structure and function. *Ecol. Model.*

Ulanowicz, R. E. and D. Baird. Forthcoming. Nutrient controls on ecosystem dynamics: The Chesapeake mesohaline community.

Ulanowicz, R. E. and B. M. Hannon. 1987. Life and the production of entropy. *Proc. R. Soc. Lond.*, ser. B, 32:181–192.

Ulanowicz, R. E. and W. M. Kemp. 1979. Toward canonical trophic aggregations. *Am. Nat.* 114 (6): 871–883.

Ulanowicz, R. E. and J. Norden. 1990. Symmetrical overhead in flow networks. *Int. J. Syst. Sci.* 21 (2): 429–437.

Ulanowicz, R. E. and W. F. Wolff. 1991. Ecosystem flow networks: Loaded dice? *Math. Biosci.* 103:45–68.

Ulanowicz, R. E. and F. Wulff. 1991. Comparing ecosystem structures: The Chesapeake Bay and the Baltic Sea. Pp. 140–166. In J. Cole, G. Lovett, and S. Findlay, eds., *Comparative Analyses of Ecosystems: Patterns, Mechanisms, and Theories.* Springer-Verlag, NY.

Van Voris, P., R. V. O'Neill, W. R. Emanuel, and H. H. Shugart. 1980. Functional complexity and functional stability. *Ecology* 61:352–360.

Varela, F. G. 1979. *Principles of Biological Autonomy.* North Holland, NY. 306 pp.

Vatican II. 1965. *Second Vatican Council: Pastoral Constitution on the Church in the Modern World.* National Catholic Welfare Conference, Washington, D.C. 138 pp.

Wagensberg, J., A. Garcia, and R. V. Sole. 1990. Connectivity and information transfer in flow networks: Two magic numbers in ecology? *Bull. Math. Biol.* 52:733–740.

Waldrop, M. M. 1992. *Complexity: The Emerging Science at the Edge of Order and Chaos.* Simon and Schuster, NY. 380 pp.

Wallace, R. L. 1978. Substrate selection by larvae of the sessile rotifer *Ptygura beauchampi. Ecology* 59:221–227.

Weigel, G. 1992. *The Final Revolution: The Resistance Church and the Collapse of Communism.* Oxford University Press, NY. 255 pp.

Westfall, R. S. 1983. *Never at Rest: A Biography of Isaac Newton.* Cambridge University Press, Cambridge. 908 pp.

——. 1993. *The Life of Isaac Newton.* Cambridge University Press, Cambridge. 328 pp.

Westra, L. 1994. *An Environmental Proposal for Ethics: The Principle of Integrity.* Rowman and Littlefield, Lanham, MD. 237 pp.

Whewell, W. 1847. *The Philosophy of the Inductive Sciences, Founded upon Their History*, vol. 2. Parker, London. 679 pp.

Wicken, J. S. 1984. On the increase in complexity in evolution. Pp. 89–112. In M. Ho and P. T. Saunders, eds., *Beyond neo-Darwinism: An Introduction to the New Evolutionary Paradigm.* Academic Press, London.

Williams, G. C. 1966. *Adaptation and Natural Selection: A Critique of Some Current Evolutionary Thought.* Princeton University Press, Princeton, NJ. 307 pp.

Wills, G. 1978. *Inventing America.* Doubleday, Garden City, NY. 398 pp.

Wilson, E. O. 1975. *Sociobiology.* Harvard University Press, Cambridge, MA. 697 pp.

Wolfenden, G. E. and J. W. Fitzpatrick. 1984. *The Florida Scrub Jay: Demography of a Cooperative-Breeding Bird.* Princeton University Press, Princeton, NJ. 406 pp.

Wolff, W. F. 1994. An individual-oriented model of a wading bird nesting colony. *Ecol. Model.* 72:75–114.

Woodwell, G. M. and H. H. Smith. 1969. *Diversity and Stability in Ecological Systems.* 22d U.S. Brookhaven Symp. Biol., NY. 264 pp.

Wulff, F. and R. E. Ulanowicz. 1989. A comparative anatomy of the Baltic Sea and Chesapeake Bay ecosystems. Pp. 232–256. In F. Wulff, J. G. Field, and K. H. Mann, eds., *Network Analysis in Marine Ecology.* Springer-Verlag, Berlin.

NAME INDEX

SUBJECT INDEX

Printed in the USA
CPSIA information can be obtained
at www.ICGtesting.com
JSHW011520221024
72172JS00014B/119